Frontispiece Oblique aerial photograph of the Werribee Gorge, Australia. This area has been uplifted by movement along a fault line which is just off the bottom of the photograph. The wide valley in the foreground has been infilled with alluvium, and the valley sides are cut in soft sedimentary rocks. The interfluves are covered with basalt lava flows. Further upstream the river runs through a gorge and has cut through harder rocks.

A wide variety of slope forms and evidence of slope processes may be seen in the photograph. The foreground slopes to the left of the river are particularly interesting. They show rounded-to-sharp breaks of slopes at the crest and an abrupt break at the base. Areas of hummocky ground indicate landsliding and there is ample evidence of gullying on poorly vegetated areas of the slope. The sharply defined, steep, straight slopes of the gorge in the middle distance are typical of rapid downcutting following uplift. (Photograph: Neville Rosengren)

SOURCES AND METHODS IN GEOGRAPHY

Series editors

M.A. Morgan PhD
Department of Geography, University of Bristol

D.J. Briggs PhD
Department of Geography, University of Sheffield

TITLES ALREADY PUBLISHED

Sediments
D.J. Briggs

Soils
D.J. Briggs

Historical Sources in Geography
M.A. Morgan

Urban Data Sources
J.R. Short

SOURCES AND METHODS IN GEOGRAPHY

Hillslope Analysis

Brian Finlayson BA, PhD
University of Oxford

Ian Statham BSc, PhD
Geotechnical Engineering Ltd

BUTTERWORTHS
London - Boston
Sydney - Wellington - Durban - Toronto

The Butterworth Group

UK
London
Butterworth & Co (Publishers) Ltd
88 Kingsway, WC2B 6AB

Australia
Sydney
Butterworths Pty Ltd
586 Pacific Highway, Chatswood, NSW 2067
also at Melbourne, Brisbane, Adelaide and Perth

South Africa
Durban
Butterworth & Co (South Africa) (Pty) Ltd
152–154 Gale Street

New Zealand
Wellington
Butterworths of New Zealand Ltd
T & W Young Building, 77–85 Customhouse Quay,
1, CPO Box 472

Canada
Toronto
Butterworth & Co (Canada) Ltd
2265 Midland Avenue
Scarborough, Ontario M1P 4SI

USA
Boston
Butterworth (Publishers) Inc
10 Tower Office Park, Woburn, Mass. 01801

First published 1980

ISBN 0 408 10622 0

© Butterworth & Co (Publishers) Ltd

Typeset by Scribe Design, Gillingham, Kent
Printed and bound by Redwood Burn Ltd, Trowbridge & Esher

British Library Cataloguing in Publication Data

Statham, Ian
   Hillslope analysis.
   1. Slopes (Physical geography)
   I. Title   II. *Finlayson, Brian*
   551.4'3   GB448   80-40564

   ISBN 0-408-10622-0

FOREWORD

During recent years, geography has been undergoing considerable change. There have been many facets to this change, but one underlying theme is the adoption of a more rigorous approach to geographical enquiry, wherever this is appropriate. It has been reflected in numerous ways: in the greater emphasis which is placed upon quantitative and statistical methods of data collection and handling; in the attention given to the study of process as opposed to the description of form in human as well as in physical geography; and in the use of an inductive rather than deductive philosophy of learning.

What this means in practical terms is that the student and teacher of geography need to be acquainted with a wide range of methods. The student, both at school and in higher education, is increasingly becoming involved in projects or classwork which include some form of individual and original research. To be equipped for this type of study he/she needs to be aware of the sources from which he/she can obtain data, the techniques he/she can use to collect this information and the approaches he/she can take to analyse it. The teacher similarly requires a pool of empirical material on which he can draw as a source of class exercises. Both must be able to tackle geographical problems in a logical and scientific fashion, to construct appropriate explanatory hypotheses, and test these hypotheses in an objective and rational manner.

The aims of this series of books are therefore to introduce a range of sources which provide information for project and classwork, and to outline some of the methods by which this material can be analysed. The main concern is with relatively simple approaches rather than more sophisticated methods.

The reader will be expected to have a basic grounding in geography, and in some of the books a working knowledge of mathematical methods is useful. The level of detail and exposition, however, is intended to make it possible for the student, with little further reading, to gain a basic understanding of the selected themes. Consequently the series will be of particular interest and use to students and teachers involved in courses in which practical and project work figure as major components. At the same time students in higher education will find the books an invaluable guide to geographical methods.

M.A. Morgan D.J. Briggs

PREFACE

Hillslopes are the basic elements of all landscapes and it is essential for the student of geomorphology to appreciate how important they are in all aspects of the discipline. Recent trends have been away from traditional landform-oriented studies to a more scientific approach where the emphasis is upon linking form to the natural processes acting over the land surface. Geomorphologists are therefore becoming increasingly aware of the contribution that physics and chemistry can make to an understanding of how a landscape develops. Hillslope studies are reflecting the current trends and so this book concentrates upon the interface between the natural sciences and hillslope geomorphology. The aim is therefore to provide a simple introduction to the theoretical background and practical experience of modern hillslope studies.

An extremely large body of knowledge has been borrowed from applied disciplines, especially civil and agricultural engineering and hydrology, and used to develop geomorphological approaches. We should not lose sight of this great contribution and we should be especially aware that practitioners in these applied disciplines have been studying hillslopes and hillslope processes scientifically for many years longer than geomorphologists. It is another aim of this book to highlight this important contribution and to demonstrate that slopes influence people's lives by constraining their activities, especially in agriculture and civil engineering. It is our belief that not enough effort is made to inform students the extent to which the natural environment can tax human resourcefulness and resources.

This book should be of value to students and teachers of 'A'-level geography, as well as to students in polytechnic and degree courses. In addition students taking courses in environmental sciences generally may find the text of relevance to their studies.

ACKNOWLEDGEMENTS

We would like to offer our sincere thanks to everyone who helped us to complete this book. In particular, Fiona Finlayson deserves great praise for stalwartly typing the draft copy. Dave Briggs was helpful too, in that he made many useful criticisms on the layout of the text. The encouraging comments from Dr Michael Morgan came at a time when the task of writing had become a little mundane, and spurred us on to the final stages of the work.

We are also grateful to the following for their kind permission to reproduce copyright material:

Zeitschrift für Geomorphologie for *Figure 6.19*;
Géotechnique for *Figure 7.13*;
Cambridge University Press for *Figure 5.17*;
Earth Surface Processes (Wiley) for *Figure 6.20*;
Quarterly Journal of Engineering Geology for *Figures 7.5, 7.6, 7.15, 7.16, 7.17*;
Capricornia for *Figure 2.4*;
Bulletin of the International Association of Scientific Hydrologists for *Figure 5.18*;
John Wiley and Sons (New York) for *Figure 3.13*.

These sources are fully referenced in the text.

CONTENTS

ILLUSTRATIONS

TABLES

CHAPTER 1   THE RELEVANCE OF HILLSLOPE STUDIES

1.1 INTRODUCTION

Hillslope analysis is a branch of geomorphology whose aims and methods are often not well understood by geographers. Yet the whole of the earth's land surface is formed of slope facets, and an understanding of their form and of the natural processes acting on them is fundamental to geomorphology. One may define a hillslope quite simply as an element of the earth's surface inclined to the horizontal. Thus a slope possesses a *gradient*, giving it a direction or *orientation* in space.

The fact that many land surfaces are inclined is very significant. Water falling as rain flows down the line of steepest slope, over the ground or within the soil, and contributes to the movement of particles of rock, soil, and dissolved material. If the slope is steep enough, debris may slide or roll without the aid of an external agent such as flowing water. These are generalised examples of *transport processes* on slopes, which form the basis of Chapters 3–5. Transport processes detach rock and soil debris from some locations (*weathering*), *transport* it some distance down the line of steepest slope, and *deposit* it again at new locations. In the course of weathering, transport and deposition, the *form* or geometry of the slopes, and so the whole earth surface, is moulded into the landscapes we see. Slope supplies the power source for these landscape-moulding processes by allowing gravity, which normally acts vertically, to act with a lateral component.

1.1.1 Some basic descriptive terms

Slopes are sites where weathering, transport and deposition take place. The study of slopes and slope processes requires careful and accurate description of slopes and this is usually done with reference to their *profile form*. A *slope profile* is a line drawn over the ground surface following the line of steepest slope. Inclination of the ground surface along that line is expressed as an angle or as a gradient (*Figure 1.1*), and its orientation as a compass bearing. The orientation of a slope is sometimes called its *aspect*. Slope profiles which have a constant angle are called *straight* or *rectilinear*; parts or *segments* of slopes may also be referred to in this way. Curved slopes, either convex or concave according to the direction of curvature (*Figure 1.1*), cannot be described by a single angle

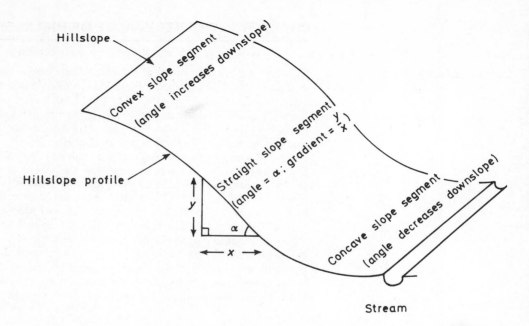

Figure 1.1 Definition of some basic
hillslope terms

but must be defined by the rate of change of angle, or *curvature*. These basic terms
provide enough background vocabulary to begin discussing hillslopes, though some techniques for measuring and analysing hillslope form are dealt with in Chapter 6.1.

Careful description of slope form is necessary in order to study its long and short term
implications. Over short time periods of a few years or so, slope angle, and to a lesser
extent overall slope form, influence the rates at which processes act and also limit the
range of processes which are possible. Over longer timespans, slope forms are actually
determined by the particular processes which have been operating for a prolonged
period. There is a tendency for a slope to reach an *equilibrium form* with continued

operation of a single process or group of processes, and therefore to become uniquely related to that process. Equilibrium forms are very interesting because once we have understood the relationship between process and form, they help us to explain landscape in terms of the processes which have produced it.

It should be said at this point that, although practically all land surface is affected by slope processes, the rate at which erosion and landscape modification is taking place is extremely variable. Some parts of the landscape are changing very slowly indeed and often retain considerable *legacy* of form due to processes which have now ceased. Much of Britain falls into this category, especially the areas of extensive glacial erosion and deposition which still retain many characteristics of a glacial landscape, partly due to the prominence of the original features and partly due to the subsequent slow rate of change. We should therefore be aware when looking at a landscape that its slopes may retain forms due to previous processes and that it is in a period of adjustment to contemporary processes. On the other hand, some of Britain's slopes are being eroded very quickly indeed and their forms are entirely related to processes currently acting on them. A good example is found in the London Clay cliffs of the North Kent coast, where the cliff-top is receding at well over 1 metre a year in some places. Here and at similar locations it is quite easy to see changes taking place from one year to the next.

1.1.2 Aims of hillslope studies

An important question to be answered about hillslope analysis is: 'What are the aims of hillslope studies and how are the aims achieved?' To the geomorphologist, hillslopes are basic landscape units and so they are fundamental to any explanation of landscape development. Thus, hillslope geomorphologists simply describe and explain hillslope evolution. How this is achieved depends on the hillslope system under observation and also on the personal approach of the researcher. The main ways in which a geomorphic problem may be tackled are discussed in Chapters 1.4 and 2.

Hillslope studies do, however, have a much wider relevance than landscape explanation alone. Civil engineers involved with major construction projects such as motorways are constantly concerned with hillslopes. Road gradients must not exceed acceptable standards and there are also important questions of slope stability, in that landsliding on embankments and cuttings should be avoided. In some areas natural processes, such as landsliding and gully erosion, present serious difficulties for construction works and need to be taken into consideration. Agriculture is also constrained by hillslopes in that mechanisation becomes more difficult, and soil erosion by flowing water over ploughed land, more acute, as slopes get steeper. It is therefore hardly surprising that civil and agricultural engineers have contributed extensively to our understanding of slopes, and this point is expanded in the remaining sections of this chapter.

1.2 CIVIL ENGINEERING AND HILLSLOPES STUDIES

It was mentioned in Chapter 1.1 that civil engineers have a keen interest in slope problems and that civil engineering has contributed a major part of the body of knowledge that has been gathered about hillslopes. In particular, much of the laboratory and field equipment used in the classification of soil materials and in the measurement of soil and rock strength (Chapter 4.2) has been developed by civil engineers, and they have also devised methods and standards of hillslope investigation which have been adopted by geomorphologists. Perhaps most important to geomorphologists is the contribution engineers have made to the understanding of processes operating on slopes, and especially the development of mechanical theories to explain the behaviour of soils and rocks on slopes. The engineer's interest in slopes is very wide indeed and ranges from questions of project feasibility and costs, through to detailed analysis and design of the project. Consequently a full discussion of his involvement in slopes would neither be possible nor appropriate here. However, an example, the theory of soil strength and mass stability, is briefly outlined in Section 1.2.1 to show how civil engineering theory can be of value to geomorphology. At the same time, the differences in approach of the two disciplines are highlighted.

1.2.1 Theory of soil strength — an example of civil engineering input to hillslope geomorphology

The strength of many materials, including soils and rocks, can be described well by a very simple empirical law called *Coulomb's Failure Law* (Chapter 3.2). This law was developed after careful experimental observation and even though it was formulated as long ago as the eighteenth century, it still forms the basis of *soil and rock mechanics*. (Soil and rock mechanics are scientific disciplines which study the behaviour of soils and rocks under stress.) It is not necessary to go into any detail about Coulomb's Failure Law here since it is fully discussed in Chapters 3.1 and 3.2; suffice it to say that it is central to the analysis of *mass-stability of slopes*, which is important in civil engineering and geomorphology alike.

Very simply, if a hillslope is **stable** there is no tendency for the soil or rock from which it is made to slide down *en masse*. Analysing slope stability is quite simple in principle, requiring the calculation of the forces tending to cause slipping and those tending to resist it, to see which are the greater. Soil strength is naturally a factor in determining the resistance to sliding; hence the importance of Coulomb's Failure Law in the analysis. Engineers use the simple ratio of *forces resisting movement to those promoting movement (the factor of safety)* to assess the risk of sliding on a slope, and if it is less than one, the slope is potentially unstable.

In civil engineering the most commonly encountered slope stability problem is to assess whether an artificial cutting or embankment will be stable when it is built (Chapter 7.1). Soil strength must therefore be measured and used to calculate a stable slope angle, using one of the many theoretical approaches which have been developed. A factor of safety in excess of one is used in the analysis of a future slope, though how much greater than one depends on the consequences of a landslide occurring after the project is complete. Thus the emphasis is on designing a stable slope on which landsliding will hopefully never occur. Conditions when the soil-body is at rest are therefore being considered (static conditions).

Many natural hillslopes are subject to landsliding and their forms may be directly controlled by the process. Thus the geomorphologist is also concerned with slope stability, and can use the theory and practice developed in civil engineering to study hillslopes controlled by landslides. The approach is in many ways similar. Soil strength is measured and used in a stability analysis to determine the critical slope angle at which sliding commences. But analysing the role of landslides in hillslope and landscape development can go much further than simply finding out what conditions initiate a landslide in a particular soil material. Unlike the designed artificial slope, landsliding has already occurred on the natural slope and so the conditions which cause the process to *cease* are also important. On geomorphic timescales, changes take place due to weathering in the soil or rock forming a slope, and can lead to changing stability conditions. Usually these long-term changes are not important to the engineer, who has to answer immediate problems of slope design, but they are significant in landscape development. The geomorphologist is also trying to answer quite different questions about slopes. Issues such as the interrelationship between process and slope form, and determination of rates of erosion are being investigated. It should not be assumed that engineers are never concerned about natural slopes. Frequently they are, because natural landslides often threaten lives and property, or are revealed in the investigation for a project and need attention. But the emphasis is still the same; a solution to an immediate problem. In conclusion one could say that, while there is a considerable body of knowledge and practice shared between geomorphology and civil engineering, the aims and approaches of the two disciplines are different. The same is true for agricultural engineering which is discussed in Chapter 1.3 below.

1.3 AGRICULTURAL ASPECTS OF SLOPES

Agriculture is almost universally carried out on soil covered slopes. All agricultural soils are susceptible to erosion at a rate which increases with slope angle and slope length. While the fundamental importance of slopes in agriculture is in their relationship to erosion, some degree of slope is important to many agricultural activities because of the drainage it provides. Some crops, notably tea and pineapples, are grown almost exclusively on relatively steep slopes because they require well drained sites. Slope characteristics are relevant to the erosion of soil by rainsplash, running water and mass movements. Wind erosion is also a major hazard to agriculture but acts independently of slope and will not be considered here. The most important problem of erosion by water is that it physically removes the soil, which is the medium upon which all agricultural activity depends. This is one major type of mechanical erosion. Some agricultural practices also lead to increased chemical erosion, especially soil drainage which increases the rate of flow of water through the soil.

1.3.1 Slope steepness and cultivation limits

Slope angles impose physical limits on the area of land that can be cultivated. Anyway, soils on steeper slopes tend to be thinner, and are therefore not well suited to cultivation. The risk of erosion increases with slope angle until it becomes too expensive to carry out the necessary works to prevent it. The cultivation of steep slopes, especially with modern heavy agricultural machinery, is both difficult and dangerous. Where agriculture is labour-intensive, for example in many south-east Asian countries, steep slopes are terraced for cultivation. In countries where agriculture is capital intensive (e.g. USA, Australia, UK etc.) terracing of steep slopes is too expensive and so steeply sloping areas are turned over to other activities such as grazing and forestry. It is interesting to note that the maximum cultivated slope angle varies from place to place. In the United Kingdom it is generally considered to be about 10°. For some areas the figures are as follows: Spain, 25°; Malaya, 18.5°; Israel, 19°; Philippines, 14°; Central Africa, 7°. Variations are partly due to differences in environmental conditions, to pressure on available land, and to differing labour costs.

Land capability classifications are frequently used to classify agricultural land according to its agricultural potential and therefore its value. Slope angle is a major factor taken into account in grading agricultural land, and is probably as important as soil properties in determining land class. There is of course a very close relationship between soil properties and slope. Not only is slope angle important but also overall profile form, plan curvature, and position with respect to the slope as a whole.

1.3.2 Agriculture and soil erosion

Agricultural scientists commonly divide soil erosion into two categories: *geological* or *natural erosion*, and *accelerated erosion. Geological erosion* occurs on undisturbed soils in their natural state and is not considered to be a problem, but an essential part of the operation of the natural landscape. This type of erosion, together with the complex group of processes involved in rock weathering, is responsible for soil formation and, over very long (geological) time periods, has produced most natural topographic features, including tracts of fertile river alluvium which have great agricultural value. Under conditions of natural erosion, soil formation and erosion are considered to be in a state of balance so that the soil maintains constant thickness through time. The natural vegetation cover is of vital importance to the maintenance of this balanced state and anything which disturbs it tends to produce accelerated erosion.

All agricultural activity involves some disturbance of the natural vegetation, ranging from its total removal and replacement by commercial crops to selective removal of some parts of the natural cover, for example timber-cropping or grazing. **Accelerated erosion** is soil loss in addition to the loss which occurs naturally by geological erosion. Since accelerated erosion is invariably not accompanied by a corresponding increase in the rate of soil formation by rock-weathering the result is a net loss of soil, often on a disastrous scale.

Accelerated erosion can be predicted by means of the **Universal Soil–Loss Equation**, devised by the United States Department of Agriculture:

$$A = R \times K \times L' \times S \times C \times P$$

where A = total soil loss per unit area, R = erosivity index, K = erodibility index, L' = slope length factor, S = slope factor, C = crop management factor and P = conservation practice factor.

This is an empirical equation which contains factors relating to the natural soils and slopes, as well as land use factors. Of central importance to this equation are the concepts of erosivity and erodibility.

Erosivity is the potential for rain to cause erosion. It is a function of the physical characteristics of the rainfall such as total intensity, drop size, and velocity. These can be combined to give a measure of the kinetic energy or total erosive power of the rainfall.

Erodibility is the vulnerability or susceptibility of a soil to erosion and is a function of both the physical characteristics of the soil and its management. While erosivity is a relatively straightforward measure of the physical parameters of rain, erodibility is much more complex because it depends on so many variables. It is important to stress that no mature, undisturbed soil exhibits any appreciable erodibility. Obviously if it did, it would have disappeared long ago. Erodibility is a soil phenomenon induced by interference, and usually by human interference. Erodibility is therefore essentially a measure of the unsuitability of a soil for the use it is, or has been, put to in that particular environment.

1.3.3 Slope geometry and surface erosion

Both the length of the slope (L) and the slope angle (α) are important in determining how much erosion will occur. Generally, steeper slopes experience greater erosion (other things being equal), since there will be more rainsplash downhill, and more runoff at higher velocities. The relationship between slope angle and erosion is not linear, but

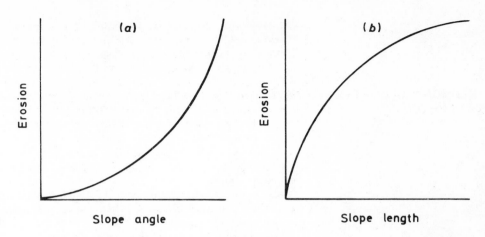

Figure 1.2 Relationships between erosion and slope angle, and slope length

has the form shown in *Figure 1.2*(*a*). This means that the amount of erosion increases *at an increasing rate* as the slope gets steeper.

As slope length increases so do the velocity and depth of runoff. It is therefore to be expected that erosion will increase with slope length. The relationship between slope length and erosion takes the form shown in *Figure 1.2*(*b*), where erosion increases at a *decreasing rate* with increasing slope length. One of the main contributors to the difference in shape of the curves in *Figure 1.2* is rainsplash erosion, which depends only on slope angle and not on slope length.

In Chapter 7.2 we will return to the problem of soil erosion in agriculture and the soil conservation measures used in combatting it. The important point to be remembered here is that hillslopes are of central importance to agriculture in terms of their relationship to soils, the limits they impose on agricultural activities, and their influence on soil erosion which makes expensive soil conservation measures necessary. Much of the basic research carried out by agricultural scientists on the mechanical erosion of soils by water, and on the analysis, description and interpretation of soils, is of direct relevance to geomorphology.

The two disciplines overlap in their subject matter but, as has already been pointed out in the case of engineering, the aims of geomorphologists are different to those of agriculturalists. In geomorphology it is natural or geological erosion which is the focus of attention.

1.4 GEOMORPHOLOGICAL ASPECTS OF SLOPES

We have seen that the disciplines of engineering and agricultural science tend to confine their attention to specific properties of a restricted range of slope types. Since virtually the whole of the landscape is made up of slopes, the study of them is the core of geomorphology. Geomorphologists are therefore interested in all kind of slopes.

Slopes can be divided into two groups on the basis of the processes which produced them. One group consists of slopes formed by constructional activities such as vulcanism and structural movements in the earth's crust. These are called internal or *endogene* processes since the force behind them is derived from within the earth's crust. Far more common are slopes produced by the external or *epigene* processes of marine, glacial, aeolian and fluvial erosion which are powered from outside the earth's crust. The most widespread slopes are those produced by the epigene processes of subaerial denudation and it is to these slopes that we will confine our attention in this book.

In fluvial landscapes, and this includes the majority of all landscapes, it is common to think of rivers as the main agents of erosion. However, in a river valley, the river itself has usually been responsible for only a small proportion of the total erosion (*Figure 1.3*).

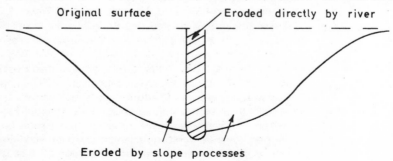

Figure 1.3 Erosion of river valleys by slope and fluvial processes

Valley widening and the development of the valley form are mainly due to the processes of subaerial denudation on hillslopes. The river's main function is to *transport* the products of erosion away from the valley sides. There are direct connections between river erosion and slope development, as the river determines the local base level to which slope processes operate. Slope development depends on the efficiency with which the river removes erosion products and the pattern of slope development will vary according to whether the river at the slope base is eroding, in equilibrium, or depositing. A river may influence the patterns of erosion and deposition on slopes by shifting its position laterally or cutting down vertically. This serves to emphasise the importance of slope processes in the operation and development of the landscape, and reinforces our assertion that the study of hillslopes is a primary function of the science of geomorphology.

1.4.1 The role of soils in hillslope geomorphology

In discussing agricultural aspects of slopes we saw that agriculturalists regard natural erosion and soil formation as being in balance with each other. Under natural conditions this must be true in the case of slopes mantled with agricultural soils. If it were not, the soil would have been eroded away long ago. In geomorphological terms these are the *equilibrium slopes*. On these slopes the soil thickness remains more or less the same while the slopes themselves may be lowered through many metres (*Figure 1.4*).

If potential erosion is greater than the weathering rate, weathered rock will be transported away before a soil can form and slope development is *weathering-limited*. The most extreme example of a weathering-limited slope is a vertical cliff, where the weathering products are transported away immediately by gravity. Although such slopes are visually distinctive and often quite impressive they really occupy only a very small proportion of the earth's surface. A third slope type occurs when the weathering rate is greater than the transport rate. These are the *transport-limited* slopes. In this case the soil cover gets thicker with time, though obviously this cannot go on indefinitely.

All slopes must show a tendency to move towards the equilibrium state. Weathering-limited slopes erode quickly and are self-destructive. Except on the most resistant rocks or where there is complete removal of material from the slope base (as in a sea cliff) steep rock slopes quickly degrade into more gentle soil- or regolith-covered ones. (Regolith is an accumulation of *in situ* weathered bedrock on slopes.) On transport-limited slopes the

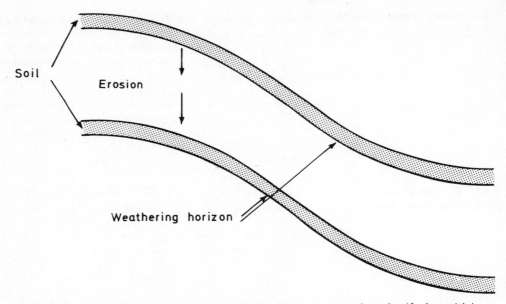

Figure 1.4 Equilibrium slope

soil cover gets deeper with time until eventually the process regulates itself, since thickening of the soil cover tends to inhibit weathering.

On soil-covered slopes the processes of weathering and erosion occur on and within the soil, and so it is necessary to understand soils in order to be able to understand slopes. Soils are the zone of interaction between the *lithosphere*, the *atmosphere* and the *biosphere*, and so they are influenced by bedrock, climate, vegetation and by the slopes on which they occur. There is a complete two-way interaction between soils and slopes, in which slope-geometry influences soil-type but at the same time soil properties, such as thickness and stability, influence slope development. These interactions are frequently studied through the *soil catena*, or soil slope sequence, which tends to be relatively constant in any particular environment (Chapter 5.3).

1.4.2 Changing approaches to hillslope studies

One of the fascinating features of slopes is their variation from place to place (spatial variation), contributing a great deal to the intrinsic beauty and interest of the natural landscape. This spatial variation is due to many interrelated factors.

A major concern in geomorphology has been, and to a lesser extent still is, description of landscape development through geological time. This is the field of *denudation chronology* and central to it is the concept of the cyclic development of landforms which will be discussed in Chapter 2.1. In this approach hillslopes are assumed to undergo a sequence of changes of form through cyclic time. One of the major problems with this approach is that the cyclic time-scale is so long that direct observation of the sequence of changes cannot be made. It has therefore been common to substitute *spatial-sequences* (changes from place to place) for *temporal-sequences* (changes through time).

Many factors other than time affect slopes and these can be divided into two groups. There are *environmental factors* such as geology, climate and climatic history, tectonic activity, and base level changes produced by fluctuations of sea-level relative to the land. These are all independent of the erosional system. Then there are factors relating to the erosional system itself, such as stream spacing (drainage density), depth of dissection or relative relief, and erosional history. Any attempt to understand slopes must take into account all of these factors and the interactions between them. Slopes are influenced by many variables, and so we cannot always rely upon spatial changes in form to be reliable indicators of changes in form at one place through time. Thus it is doubtful if cyclic ideas really aid correct interpretation of landscape development (Chapter 2.1).

More recently, attention in geomorphology has shifted from the development of slopes through time to the interrelationships between form and process. This has necessitated the development of techniques for measuring hillslopes and expressing their forms in quantitative terms, both in cross-section and in plan (Chapter 6.1). It has also brought with it a need to measure processes operating on hillslopes, and a need for the development of models of hillslope development based on the interrelationships between form and process (process–response models). These developments have led to a change in the way observed slopes are interpreted. Explanation now tends to be in terms of force and resistance and the interactions between materials, forms and processes rather than in terms of an assumed sequence of development through cyclic time. One of the most important results of modern quantitatiye work on slopes has been to reject the idea of an all-embracing theory of slope development and to emphasise the great variety of slope-forms and processes acting on them.

As we have said earlier, the purpose of hillslope geomorphology is the description and explanation of hillslope evolution. There is no single correct way to fulfil this purpose and geomorphologists use a great variety of skills and techniques and hold many different points of view on the subject. Some of the common ground with engineering and agriculture has already been mentioned and in later chapters we will see that in order to be able to understand hillslopes we must delve into subject areas such as physics, chemistry, soil science and hydrology.

CHAPTER 2 APPROACHES TO HILLSLOPE STUDIES

2.1 HISTORICAL BACKGROUND

In Chapter 1 we mentioned that a number of different approaches to hillslope problems are possible, and this section concentrates on the historical development of hillslope geomorphology. Over the last 100 years or so, geomorphologists have been striving to achieve broadly similar aims; that is, to explain the evolution of hillslope form through time. The way that this aim has been tackled has, however, changed dramatically as geomorphology has developed.

2.1.1 The nineteenth century

Many early geomorphologists were keenly aware of slope processes and made many perceptive observations on their operation. Much of the early theoretical and experimental work on soil creep, for example, was carried out in the mid-nineteenth century by a geomorphologist called Charles Davison. G.K. Gilbert, another eminent geomorphologist of the late nineteenth and early twentieth centuries, wrote about the influence of rainsplash upon hillslope form in semi-arid areas. Some of the early geomorphological work was therefore orientated towards explaining landscape in terms of the processes that formed it and, while few measurements on rates and mechanics of process were made, much pertinent observational work was carried out.

2.1.2 The cyclic–evolutionary theories

The end of the nineteenth century saw the increasing dominance of geomorphology by quite a different approach; the cyclic–evolutionary theories of landscape development. (Cyclic because the theories assume that slope development occurs in continuous cycles of tectonic uplift and erosion; evolutionary because slope development follows fixed trends, with one form always predetermined by the previous one in the sequence.)

The underlying principle of this approach is that landmasses are uplifted rapidly by tectonic activity, and are then subjected to weathering and transport to reduce them steadily towards a low plain close to sea level. A fundamental concept is that of *base level*, below which no erosion or transport can effectively occur. On the continental scale, base level is simply sea level, though locally it may be controlled by some other level, such as a lake. Also important is the assumption that rivers rapidly achieve a *graded profile* when downcutting into a land surface. The concept of grade is rather difficult to explain in detail, but as seen by the early proponents of cyclic–evolutionary models it can be taken to mean a smooth profile, roughly in balance in terms of the material being delivered to it and carried away by the river (*Figure 2.1*).

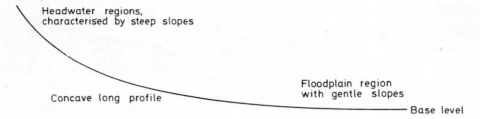

Headwater regions, characterised by steep slopes

Concave long profile

Floodplain region with gentle slopes

Base level

Figure 2.1 The traditional view of a graded profile

W.M. Davis was the main proponent of the cyclic–evolutionary approach. Broadly speaking, he envisaged landscapes following the trend outlined below. Rapid tectonic uplift takes place at such a rate that erosion is insignificant during this early phase. After uplift, erosion commences with very rapid downcutting of streams into the new surface. This leads to the establishment of very steep-sided valleys. When the streams attain a graded profile, vertical erosion practically ceases and the hillslopes begin to decline in angle under the influence of the agents of erosion (wearing down). Finally, lateral erosion by the river begins, leading to the development of an extensive floodplain and even gentler slopes. The ultimate landscape is the *peneplain*, a low, barely undulating surface with sluggish, meandering rivers. The cycle may be restarted with a new period of uplift. It should be stressed that Davis wrote many essays on landscape development and the brief outline above is only the skeleton of his ideas about humid landscapes (illustrated in *Figure 2.2*).

A German geomorphologist of the day, W. Penck, criticised the basic premise of the Davisian theory. He suggested that uplift need not necessarily be faster than erosion and that the latter could keep pace with the former, in some cases. The balance between the rate of uplift and rate of erosion affects the form of hillslopes, as envisaged by Penck and

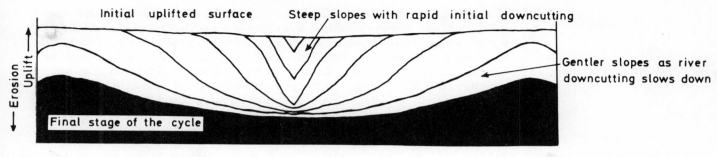

Figure 2.2 An example of the Davisian approach to cyclic slope evolution — the humid cycle

Davis, occurring throughout the cycle of erosion. Interestingly enough, recent precise measurements of uplift in tectonic regions and erosion in mountainous districts, indicate that uplift is very much faster than erosion, which supports Davis' original assumption.

2.1.3 Denudation chronology

Davis' model of landscape development considers hillslope form to be indicative of the length of time the slope has been exposed to erosion, and terms such as 'youthful',

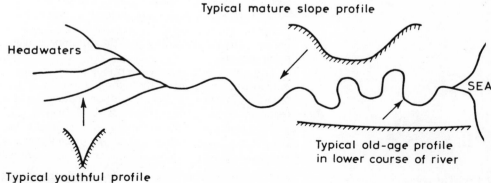

Figure 2.3 Davis' concept of the age of slopes

'mature', and 'old age' loom large in his discussion. Quite simply, he considered the age of a landscape to be the basic control on its form (*Figure 2.3*). These ideas carried through into **denudation chronology**, which was a popular way of analysing landscape in the first half of the twentieth century. Denudation chronology is concerned with elucidating the different phases of development that can be seen in a landscape. These phases are assumed to relate to base level changes, usually due to minor changes in sea level. If sea level falls with respect to the land the rivers will be able to erode lower, and so a new phase of rapid downcutting to re-establish the graded profile will begin. A rise in sea level causes the profile to be adjusted by deposition. Changes in sea level are often equated with the development of suites of river terraces.

2.1.4 Drawbacks of the Davisian approach

The cyclic–evolutionary theory of landscape development may be broadly correct as far as it goes. Indeed, the suggestion that landscapes become more subdued as time goes on is a very reasonable one. But this basic approach has been very heavily criticised recently for a number of reasons. Undoubtedly the most serious flaw in Davis' work is the total lack of any substantive data (or even accurately reported qualitative observation) to support his theories. Also, he totally disregards the influence of slope processes on slope form. Processes of weathering and transport are always referred to vaguely as agents of erosion and there is never any suggestion that different slope processes might lead to different forms. True, different erosion cycles were proposed for different climatic regimes, but the possibility that the differences might be due to process variation was never explored.

Davis also made gross assumptions about the age of parts of the landscape. His sequences of slope forms through *time* were assembled from observations made from slopes separated in *space*, on the assumption that slopes become progressively older as one proceeds down river valleys (*Figure 2.3* and Chapter 1.4). It was never established that age and position were quite so simply related. Finally, Davis' statements, drawings, and interpretations of other people's ideas, are often vague and contradictory. Consequently they do not really help us to explain slope forms in a satisfactory way.

2.1.5 The revival of process studies

Over the past 20 years or so there has been a drift away from cyclic theories of slope evolution and a revival of interest in slope processes. This major shift in emphasis was probably inspired by the failure of cyclic models to explain anything other than broad trends in landscape development. It is not possible to answer questions about how a specific slope reached its current form in terms of Davisian geomorphology. Also, there was an increasing awareness that there was a need to understand the operation of slope processes in order to understand the hillslope system as a whole.

The aim of process studies is to explain relationships between process and form in order to understand the current state of development of slopes. There are two main aspects of processes which relate to their influence on form. First, the mechanics of a process, or *how* it operates, and secondly the measurement of the *rate* at which it operates along the slope profile, which controls the way form is changing through time. By studying these two aspects we can begin to understand the hillslope system more clearly, and hopefully avoid prejudicing our conclusions on how slopes develop through time.

A vast amount of data has now been gathered on processes, though often the emphasis is too much on the process in isolation and not enough on its influence on form and slope development. Perhaps because of this there has been a very recent revival of interest in denudation chronology. The difference between this recent denudation chronology and earlier work is that a much more rigorous approach has been possible. The importance of processes is now recognised and remnants of past phases of erosion can be explained in these terms. Also, modern dating techniques (such as radiocarbon dating) allow absolute ages to be placed on parts of the landscape. This is a very promising development for hillslopes since the ability to put a true age on slopes may help to produce general theories of slope development which do not suffer from the subjectivity of earlier work.

2.2 THE ROLE OF THEORY IN HILLSLOPE STUDIES

Any scientific study relying upon observation as its source of information requires theory to transform bald observational facts into an explanation of how the system operates. A theory can be regarded as a statement or set of statements designed to link and explain a set of observations. There are two ways in a scientific study in which a theory may be generated. It may be derived *deductively*, that is the theory is generated from an initial idea from which all other statements in the theory follow. In this case, theory is generated and then observations are made to substantiate it or refute it. Secondly, theory may be derived *inductively*, in that a series of observations already made may lead to the generation of a theory in order to explain them. In practice, both deductive and inductive approaches are used in most scientific studies.

A theory need not be a complex mathematical argument but could be a simple speculative idea occurring to someone in the process of gathering data, or even prior to studying a problem seriously. But all theories possess broadly similar characteristics. First there are *initial postulates*; the ideas and assumptions upon which the theory is based. These are subjective statements, quite indefensible in strict logical terms, and must be taken as true statements within the bounds of the theory itself. For example, one of Davis' basic postulates is that erosion is insignificant during the tectonic uplift stage of the landscape development cycle. Although the basic postulates are subjective statements they are very important indeed since they represent a sideways jump from straightforward logical thought to a fresh and original line of reasoning. Following the basic postulates are *derived statements* which follow logically from the initial ideas. In practice there are three main types of theory useful in geomorphology: speculative or intuitive theory, theories from classical mechanics, and statistical theories. These categories are broadly described below.

2.2.1 Types of theory

Intuitive or speculative theories are simply ideas which spring to mind about how something operates. They may be based on considerable observation or upon very little, but they are always essentially embryonic in that their arguments are poorly formulated or cannot easily be tested. Thus, speculative theories should be the first step towards developing a scientific theory. Cyclic theories of landscape development (Chapter 2.1) barely go beyond the bounds of speculation because, although considerable argument follows from the basic postulates, it is developed in such a way that testing is virtually impossible. For this reason, this type of theory will not be referred to again in this book.

Theories from classical mechanics form the second category of theories important in hillslope geomorphology. In order to understand a hillslope process and its impact on slope form we need to know *how* it operates, and this is often easily done in terms of established natural sciences. In many cases it is possible to develop a strict mechanical theory of the process which can then be applied to real world data with some adjustments. In order to do this, one must make simplifying assumptions about the whole system under study, since natural processes and materials are so very variable. Accordingly we should expect that the data we collect about the real world will have a degree of 'randomness' (randomness is used here in a loose sense to mean variability introduced by factors we neither recognise nor understand) not predicted by theory, and the relationships will not be so clear as the theoretical ones.

Perhaps an example will make the point clearer here. In the accumulation of *scree slopes*, slopes made up of boulders which have fallen from a cliff and accumulated at its base (Chapter 6.2.7), we can model the process of slope development in terms of the simple process of single rocks falling from the cliff through space, to accumulate on the scree below. In order to understand the process and to show how it causes scree slope form to develop through time, it is necessary to predict how far pebbles falling from any height on the cliff will travel down the scree before coming to rest.

This can be done very easily indeed using the standard mathematical equations of motion to calculate impact velocity of the pebble and the downslope travel distance on the scree. But even in this very simple case, many assumptions need to be made in order to make the calculations feasible. For example, one must assume negligible air resistance to calculate pebble impact velocity, that pebbles fall freely and do not bounce down the cliff, and that they slide over the scree surface rather than roll or bound, and so on.

We find that if we conduct experiments to test the simple theory we discover that there is considerable scatter upon the data, though a relationship similar to that predicted by the theory is generally found. (In 6.2.7 this simple theory is outlined in more detail.) According to the theory, every pebble falling from the same height would travel exactly the same distance down the scree slope. The scatter on the real world data shows that this is not entirely true, and indicates that some of the assumptions made do not always hold. The scatter is due to many factors which are not easily described in strict mechanical terms and it emphasises that there will always be variations in any natural system which cannot be accommodated in a strict mechanical theory.

Statistical theories form the final type of theory useful in hillslope geomorphology.

Since natural systems are never rigid in operation, statistical approaches can often be used as an alternative to classical mechanics in order to describe and explain processes or the development of form. One brief example of how a statistical approach may be an alternative to mechanics is given here; that of the initiation of sediment transport in water flowing over a slope or in a channel. Using mechanics, one could calculate the forces tending to cause movement, due to the water flow, and those tending to resist movement, due to factors such as the weight of the grains, and erosion would begin when the forces tending to cause movement exceeded those tending to resist it.

This approach is acceptable theoretically, in that the argument is irrefutable, but since the basic postulates are too rigid there is considerable lack of agreement between theoretical predictions and real observations. The main problem is that it is necessary to assume that the force imposed by the flowing water is steady and evenly distributed. This is far from the case, and most transport is probably initiated in constantly changing swirls or *eddies* which concentrate force at some points. An alternative is to view the initiation of sediment transport at any point in terms of the *probability* of erosion occurring at that point in a given timespan.

Probability theory is a branch of statistical theory with a considerable degree of mathematical sophistication. Put simply, it means that we can largely dispense with classical mechanics and explain the process in terms of the chance of erosion occurring. The factors affecting that chance (or probability) are still complex, however, and require some consideration of mechanics as well. There are many other ways in which statistics may be useful in hillslope studies. They can, for example, be used to isolate so-called variation, or to demonstrate links between factors not immediately obvious. Both of these can be of immense value in interpreting how a system operates.

2.3 THE ROLE OF MEASUREMENT AND EXPERIMENT

As we have just seen, theory is important to the study of hillslopes and without a continually developing theoretical approach to the subject it would not progress. However the theoretical approach alone is not sufficient. Theories must be tested by measurements, observations and experiments, and as the results of practical work become available theories are modified in the light of the new information, and new theories are developed. Thus measurement and experiment in the field and in the laboratory are essential for progress in hillslope studies. While we stress the importance of quantitative work throughout this book, detailed and accurate qualitative observations also have a place in the study of hillslopes. When working on hillslopes in the field, or when conducting experiments in a laboratory, it is essential to observe carefully and keep detailed notes. The ability to observe properly comes only with practice.

In the study of hillslopes, geomorphologists use a variety of techniques and approaches, many of them shared with other disciplines. A necessary prerequisite to all slope studies is the measurement of slope form both in plan and cross-section (Chapter 6.1). It is also necessary to describe and analyse the materials forming the slopes. This includes field descriptions and laboratory analyses of both soil and bedrock. In the laboratory, measurements are made of the properties of materials such as strength, grain size distribution, and chemical composition. Measurements are also made of the processes operating on slopes to determine how fast individual processes work, how fast whole slopes are eroding and to discover exactly how individual processes operate. However, hillslope processes are very complex and it is often difficult to sort out, under field conditions, the relationships between the many factors which influence them. For this reason it is often helpful to eliminate some of this complexity by setting up controlled experiments in a laboratory or on an experimental slope plot.

2.3.1 Slope form on maps and in the field

One of the most remarkable things about much of the early geomorphological work on hillslopes is the almost total absence of any measurements of slope form. As pointed out in Chapter 2.1, many very elaborate discussions were carried on by early geomorphologists about slope forms and the manner of their evolution without a single measurement being made. Given the ease with which slope form can be measured (Chapter 6.1) this omission is surprising. Form can be described both in plan and cross-section (*Figure 2.4*).

Maps and air photographs
The most obvious sources of information about slope form are published contour maps. The limitations imposed by the scale of the map and the contour interval mean that they are usually used only for reconnaissance surveys or for mapping fairly generalised slope

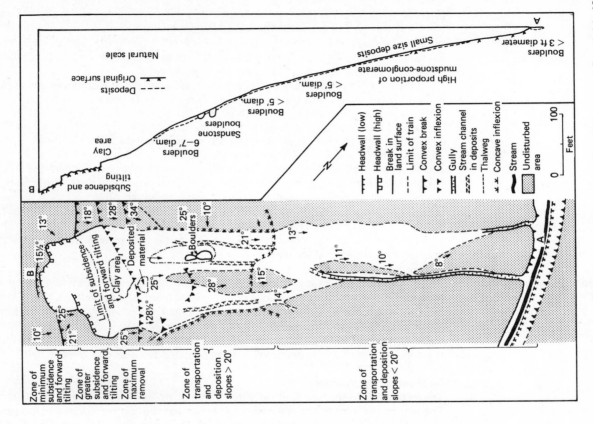

Figure 2.4 Morphological map and hillslope profile (After Finlayson and Greenhalgh, 1969)

classes, as in the land capability surveys mentioned in Chapter 1.3. The contour intervals used on published maps are so large that most of the interesting detail is obliterated. Some of this detail can be seen on aerial photographs, but it is very difficult to extract precise quantitative information on slope from them. In hillslopes studies, published maps and aerial photographs are used in the planning stage and to construct base maps for use in the field.

Field mapping

Field mapping is undertaken for a wide variety of purposes. It is common practice to construct detailed contour maps with a contour interval as small as 10 cm for irrigation, soil conservation, road building and other civil engineering projects as well as for geomorphological research projects. Geomorphologists have for many years made *morphological maps* (*Figure 2.4* and Chapter 6.1) as part of their investigations on slope forms in plan and these maps are now also being made for engineering purposes, such as highway design and landslide control. A morphological map uses symbols to portray the configuration of the land surface and it is possible to map features in this way which would lie within the contour interval on a contour map. Morphological mapping is as much an art as a science, and requires sensitivity on the part of the map maker to subtle changes of topography, which can only be acquired by practice.

As we have already seen, gravity is the major force operating on hillslopes and it acts down the line of steepest slope. This line is the *hillslope profile* (*Figures 1.1, 2.4* and Chapter 6.1) and crosses all the contours at right angles. Rivers also cross all the contours at right angles, and so a river channel is simply a rather specialised hillslope profile. Measurement of hillslope profiles is the geomorphologist's most important technique for studying slope form and we will consider it in more detail in Chapter 6.1. The profile is used as the basis for studying form–process interrelationships on hillslopes (Chapter 1.1).

Detailed measurements of processes operating on hillslopes is a relatively recent development in geomorphology and a great deal of work remains to be done in this important area. We should not forget, however, that during the nineteenth century a lot of process observations were made but the appearance of cyclic models of landform development around the turn of the century shifted attention to form, and it is only recently that lines of research begun in the last century have been picked up again. Perhaps we would now be much further advanced in our understanding of hillslopes and of the whole field of landscape development if we had continued as the nineteenth century pioneers began, with a lot of healthy curiosity and no all-embracing cyclic models to constrain our thinking.

2.3.2 Process measurements

Process studies are aimed at measuring how fast individual processes operate; at understanding the mechanics of those processes; at measuring rates of hillslope erosion; and at developing models of the interactions between form and process on hillslopes. Some processes operate more or less continuously over the whole of a hillslope while others occur sporadically only on certain types of slopes. For example, in humid climates the removal of material in solution goes on fairly continuously over the whole of the slope, especially where a soil cover is present. Soil creep is also a widespread process but is more seasonal in its activity. Landslides and other forms of rapid mass movement occur only on a restricted range of slope types and many years may elapse between one movement and the next. Slope processes vary through time and each process has its own individual pattern of *temporal variation*. They also vary through space and there are individual patterns of *spatial variation* for each process. Understanding these patterns for any one process is a difficult enough task but we must also try to understand how they all fit together. It is very difficult to compare the rates of processes having different temporal patterns especially since there are usually only short records of their operation available. For this reason, measurements of processes must be evaluated in the light of what can be learned about a slope from its form and the materials on it.

Many hillslope processes, for example soil creep and chemical weathering, operate very slowly and are therefore difficult to measure. There are a number of ways of overcoming this problem. Observations can be made over a long time period (e.g. 10 years) so that enough will have happened in that period to be detectable. Alternatively, short period movements may be magnified mechanically or detected using instruments capable of high resolution. This leads the geomorphologist into the field of instrumentation and a lot of research is currently being carried out to develop instruments to measure slope processes over short time periods. We will learn more of these techniques in later chapters.

2.3.3 Laboratory experimentation

If the mechanics of processes are to be fully understood the properties and states of the materials involved must be known. Also, measurements made on materials can be used in the prediction of slope processes. For example, measurements of strength made on the soil and rock which has moved in a landslide help in reconstructing the circumstances which caused it to occur, and strength testing is also necessary for predicting the conditions under which failure will occur on any particular slope. In fact, most field process studies are accompanied by laboratory analyses of the materials involved.

Yet another approach is to reproduce the process in a laboratory and speed it up so that many years of operation at natural rates are condensed into much shorter time periods. Clearly it is not possible to reconstruct a whole hillslope in a laboratory, and in any case no useful purpose would be served by this since the aim of laboratory simulation is to simplify reality so that it can be more easily understood. As we have already mentioned in Chapter 2.2.1, one possible exception to this is a scree slope, which can be reproduced in a scaled down version in a laboratory.

The question always remains of how well the scaled down version reproduces reality. For example, do the small pebbles used in laboratory screes behave in the same way as large boulders on real screes? And this is only one of a host of limiting assumptions made in producing a laboratory model of a scree slope.

Most laboratory simulation is aimed at reproducing just one process or even just one aspect of a process. The complexity of natural hillslopes is deliberately simplified to study the effects of selected variables. Laboratory simulations that have been made include chemical and mechanical weathering, soil creep, rainsplash and the flow of water through soils. Some workers have even produced soil profiles in a laboratory by subjecting rock particles to accelerated chemical weathering. Laboratory simulation is fraught with difficulties and hidden traps. A change of scale is usually necessary and its effects are often not fully understood. It is difficult to model the parameters and the materials correctly and this can lead to spurious results being accepted as correct.

2.3.4 Griggs' rock-weathering experiments — a case history in laboratory simulation (Griggs, 1936)

A good example of laboratory simulation of slope processes is provided by the work of the geologist, David Griggs, during the 1930s. It has long been observed that rocks break up into small pieces in hot deserts and it was assumed that this was due to stresses set up in the rock by heating during the day and rapid cooling at night. Griggs attempted to reproduce this process in a laboratory by subjecting a block of granite to heating and cooling cycles, each cycle lasting fifteen minutes. By speeding up the process in this way he was able to reproduce the equivalent of 89 400 daily temperature cycles which would have taken 244 years under natural conditions. The granite did not shatter and no changes could be detected, even under the microscope. However, he subsequently found that if he cooled the granite with a spray of water instead of with dry air, visible changes occurred after the equivalent of only two and a half years of natural weathering. He concluded therefore that mechanical shattering of rocks is also partly a chemical process since water is essential for its operation.

Griggs' experiment is instructive. It shows us the importance of questioning commonly held views and testing their validity. It also demonstrates that rejecting a hypothesis is as important a step in scientific discovery as accepting one. When Griggs' rock did not disintegrate in the dry tests the experiment was not a failure, it was a success. The experimental technique was quite simple (perhaps the reader would like to reproduce it for himself) and shows that much can be learned from simple laboratory experiments. The question does remain, however, of whether or not a series of fifteen minute cycles is a good enough imitation of a series of 24 hour cycles. Also, the effect of taking a small sample is not known. In short, did Griggs model the parameters of the process correctly? We do not really know the answer to that question, but the important thing when observing, measuring and experimenting is to keep an open mind, take nothing for granted and keep asking questions.

CHAPTER 3 THE APPLICATION OF PHYSICAL SCIENCES TO HILLSLOPE STUDIES

In Chapter 2 a number of ways in which one might study hillslopes were discussed. The approach taken in the remainder of this book, which was outlined in Chapter 1, is based on an analysis of the processes of sediment and solute movement on slopes, and the relationships between these processes and the development of hillslope form. Movement of material implies that some sort of force is brought to bear upon it; which may be mechanical or chemical. This of course implies that hillslope geomorphologists can learn from physics and chemistry and use their basic concepts. In this chapter we shall try to show how an appreciation of simple mechanics and chemistry can lead to a better understanding of how hillslope processes operate. Indeed, not only does this lead to an increase in understanding, it also allows theories of process-action to be postulated which can then be tested in the laboratory and field. Much of this chapter will be familiar to those who have a background in physics and chemistry, though the relevance of the material to hillslope processes is stressed throughout.

3.1 FORCE AND RESISTANCE IN HILLSLOPE PROCESSES

Hillslopes are usually covered with some sort of sediment. This may be soil or regolith derived from weathering and breakdown of the bedrock in which the slope is cut, or it may be material which has been derived elsewhere and transported to its current position. For the most part we will be concerned with *soils*, that is material derived by weathering from the bedrock underlying the slope. The form of nearly all slopes is controlled by where soil is and how much there is of it, and also by the way in which it is moved and how fast. Even in steep mountainous regions bare rock slopes are much less common than you would imagine and over the earth as a whole they are rare indeed. Consequently when we look at a landscape we are, by and large, looking at hillslopes covered in soil which is involved in processes of weathering and transport. In order to understand how this soil is transported, it is necessary to look at some simple mechanics.

3.1.1 Force and source of force on hillslopes

All movement requires the application of a **force**. A force is an action which tends to change the state of motion of a body. More simply, if you apply a force to a stationary object you will tend to make it move in the direction of the applied force, or if it is already moving you will tend to make it change speed or direction (or both). It is important to note that there is only a **tendency** for the state of motion to change. Everyone must have experienced the frustration of exerting great effort in trying to lift something, and yet not moving it at all!

In geomorphology, forces are derived from only a few sources, the most important of which is **gravity**. Gravity is a mutual attraction between bodies and this means that objects on earth's surface tend to be pulled more or less vertically towards the centre of the earth. The force exerted upon an object resting on the earth's surface can be taken as equal to its weight. This is something of an oversimplification however, and it certainly does not apply to objects which are accelerating or decelerating. This point will be returned to later.

Gravity is often directly responsible for the movement of sediment particles. For example, a particle of rock detached by weathering from a cliff will fall vertically, under the influence of gravity, to the ground. Perhaps more often though, gravity acts through some other agency, the best example of which is flowing water. Gravity causes bodies of water to move on slopes which in turn exert forces on sediment particles, tending to transport them. An extreme example of gravity acting upon water to cause sediment movement is the impact of raindrops on bare ground, which sends soil particles flying into the air (Chapters 5.2 and 6.3).

Water is capable of exerting other forces upon and within soils, which are important in transport processes. Water contained in the pore-spaces of a soil behaves in a number of ways. If the pore-spaces are not completely filled with water (i.e. the soil is **unsaturated**; *Figure 3.1*) then a suction force is exerted which tends to draw the soil grains more strongly together. This suction is called **capillary tension** and everyone will have experienced its importance on the beach as the force which makes sandcastles stand up. You will have all noticed that you cannot build sandcastles from completely dry or completely saturated sand; it needs to be damp to take advantage of the capillary tension forces. If the pore-spaces are completely water-filled (*Figure 3.1*), that is the soil is

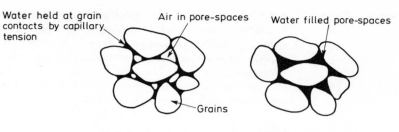

Water held at grain contacts by capillary tension

Air in pore-spaces

Water filled pore-spaces

Grains

(a) Unsaturated   (b) Saturated

Figure 3.1 Saturated and unsaturated sediments

saturated, the water exerts a pressure within the pore-spaces which tends to push grains apart. The precise way in which this pressure acts and its magnitude are dependent on many factors which are too involved to be elucidated here. You may get a fuller under- standing of how water pressure forces operate in soils from the experiment unit below (Chapter 3.1.2).

A third major source of force on hillslopes is due to expansion and contraction of soil grains, or of water within pore-spaces and cracks in bedrock and regolith. Expansion cycles may be caused by a number of mechanisms, all of which are related to climatic factors. Because of this they are often called *climatic forces*.

Direct heating by the sun and cooling at night is a major cause of expansion and con- traction. Soil particles or fragments of rock tend to expand and contract preferentially down a slope, causing a net shift in position throughout each temperature cycle. The mechanism is discussed more fully in Chapter 4.4.

Water contained in cracks or pore-spaces in rocks and soils expands on freezing by about 9% and contracts again when it melts. This again may cause net shift of material.

Finally, and perhaps the most important mechanism on a world scale, many soils contain appreciable quantities of *clay minerals*, some of which are able to swell and contract as their moisture content increases and decreases respectively. They are able to do this because of their crystal structure, which is a series of alternating plates composed of silica (SiO_2) and aluminium hydroxide ($Al(OH)_3$). This is discussed later in Chapter

3.2.2. Some clay minerals can absorb considerable quantities of water between plates, and so their crystal size is highly variable. The process of swelling and contracting with soil moisture change can again be responsible for downslope movement of material. The role of soil moisture change and freezing in soil creep processes is also discussed in Chapter 4.4.

Plants and animals are able to exert quite large forces on soils and can be locally very important. Their effects range from slow pushes due to root growth, upward movements

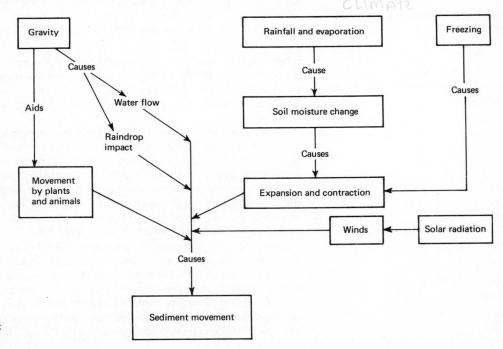

Figure 3.2 Forces involved in sediment movement (After Statham, 1977)

of material by smaller organisms such as ants and worms, to larger scale movements by burrowing animals such as rabbits. Some animals are also important not because they directly move soils but because they remove the vegetation cover. Later we shall stress the great importance of vegetation throughout hillslope processes in general. This is especially brought out through the whole of Chapter 5 where water on hillslopes is discussed.

To summarise what has been said in this introduction to forces on hillslopes; a major source of force is due to gravity acting vertically upon soil and upon water. Other important sources are due to water pressure forces, lateral expansion and contractions related to climate, and forces exerted by plants and animals. These sources of force are illustrated in *Figure 3.2.*

3.1.2 Experiments to demonstrate water pressure and suction forces in sediment

These experiments show how water pressures can be important in sediment transport.

3.1.2(a) Water pressure exerted upon a pebble submerged in water

A pebble weighs less in water and therefore is more easily transported by flowing water. Attach a pebble to a spring balance and record its weight in air. Lower the pebble (still attached to the spring balance) into a graduated cylinder of water and record the water level in the cylinder both before and after lowering the pebble into it. Record also the weight of the pebble when suspended in water. Notice that the pebble weighs less in water and that it weighs less by an amount equal to the weight of water displaced by the pebble (1 g of water has a volume of 1 cm^3).

Specimen Results

1.	Weight of pebble in air	295.0 g
2.	Weight of pebble in water	173.7 g
3.	Initial volume of water	700 cm^3
4.	Volume of water with pebble	811.3 cm^3
∴	Volume of water displaced	111.3 cm^3

The upthrust force on the pebble is equal to the weight of water displaced which, in this case, is 111.3 cm³. Since:

$$\frac{\text{force of buoyancy}}{\text{weight of pebble in air}} = \frac{\text{density of water}}{\text{density of pebble}}$$

$$\text{density of pebble} = \frac{\text{weight of pebble in air} \times \text{density of water}}{\text{force of buoyancy}}$$

The density of water is $1\,\text{g cm}^{-3}$ so that:

$$\text{density of pebble} = \frac{\text{weight of pebble in air}}{\text{weight of water displaced}}$$

In the case of the specimen results shown above the density of the pebble is $2.65\,\text{g cm}^{-3}$.

3.1.2(b) Water pressure forces within a sediment

Water can exert forces in sediment grains which are sometimes sufficient to push the grains completely apart, resulting in complete loss of strength.

Connect a burette filled with water to a container filled with coarse sand as shown in **Figure 3.3**. Add water to the sand until it is completely filled. The water level in the burette will be the same as that in the sand. Close the tap and place a heavy object (e.g. a large iron bolt) in the sand. Add water to the burette until it stands three times the thickness of the sand above the sand surface level. When the tap is opened the water flows upwards through the sand causing it to bell upwards. At the same time the iron bolt will sink into the sand. Another similar bolt added after the flow has stopped will not sink.

The pressure exerted by the water flowing upwards through the sand was sufficient to force the grains completely apart, resulting in a complete loss of strength, and the sand became a quicksand. The pressure exerted by the water is proportional to the height of

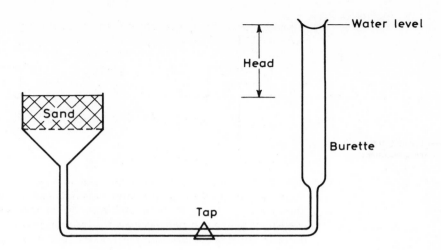

Figure 3.3 Experimental setup for 3.1.2(b)

the water column (head) in the burette above the level of the sand surface. You can demonstrate this by repeating the experiment using lower heads. Flow will still occur but the bolt will not sink.

3.1.3 The mechanics of soil movement and resistance to movement

It has already been stated that a force is an action tending to change the state of motion of a 'body'. Most of the 'bodies' with which we are concerned are soil particles or masses of soil moving as a single unit and subsequent discussions will be in these terms. A most important concept to grasp about force systems is that they are always in equilibrium. That is, if a force is exerted on a body there will be an equal and opposite force set up called the *reaction.*

Let us consider an example of a force system which demonstrates that the reaction is a real force and not one required conceptually to complete a force equilibrium. Imagine a boulder of weight *W* acting vertically on the ground surface (*Figure 3.4*(a)). The weight of the boulder tends to make it sink into the ground and the fact that it does not do so

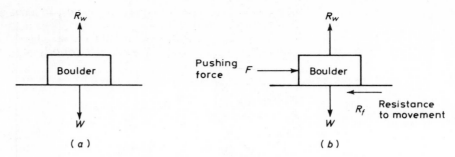

Figure 3.4 Force systems acting on stationary and moving boulders

means that it must be resisted by an equal and opposite reaction (R_w) within the ground. In this static case the reaction is the force maintaining the boulder in a stationary condition and we might call this a *static equilibrium of forces*. In fact, even if the boulder were moving at a constant velocity, the reaction (R_f) will exactly balance the pushing force (F) (*Figure 3.4(b)*). This reaction is due to frictional resistance between the boulder and surface, which is discussed in 3.1.6. In the case of a boulder which is accelerating or decelerating, the reaction force is not so simply envisaged. However, since such situations are difficult to analyse mechanically and are not usually important in terms of geomorphic processes, they will be discussed no further.

3.1.4 Force resolution and combination

Hillslope process studies almost always require the analysis of forces acting on particles or masses of soil on sloping surfaces, which are tending to cause movement downslope. The main force, that due to gravity, does not operate directly down the slope but acts vertically. Consequently it is necessary to find *how much* of the gravitational force acts in the direction of slope, which involves a procedure in mechanics called the *resolution* of forces. It is possible for *components* of a single force to operate in two or more directions and very simple rules exist to find the magnitude of each component. This is best illustrated by the simple example of a boulder resting on a slope (*Figure 3.5*).

The force due to its weight acts vertically and is balanced by a reaction force (R_w). We are interested in force components acting on the boulder in two directions; parallel to the slope which tends to cause the boulder to move downslope, and normal to the slope which tends to keep the boulder static. The magnitude of these force components may be found by constructing a parallelogram where OC is a diagonal of the parallelogram and is drawn proportional to the weight of the boulder (*Figure 3.5*). OA and OB are the

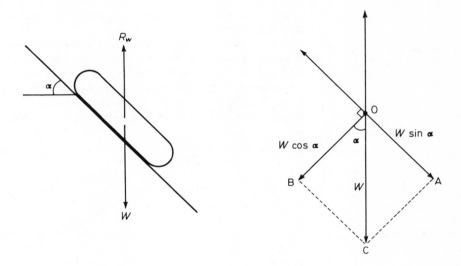

Figure 3.5 Forces acting on a boulder resting on a slope

components of the weight **W** acting parallel and normal to the slope respectively, and their lengths are drawn such that all internal angles in the parallelogram are right angles (that is, in this case the parallelogram is a rectangle, because the directions of the components are at right angles to each other). If the diagram is drawn accurately, the lengths of OA and OB are proportional to the magnitude of the force components, which can be found by measuring them.

This is a clumsy way of finding their magnitudes, and in practice this is best achieved by simple trigonometry. Returning to *Figure 3.5*, it can be seen that

$$OA = W \sin \alpha$$
$$OB = W \cos \alpha$$

which provides a quick and easy way for determining the forces acting on the boulder in different directions.

It is possible for two or more forces to **combine** their effects on an object and to act **in a single direction**. A good example of combination of forces might be provided by a sand grain carried along by a fast flowing river (*Figure 3.6*). The submerged weight (**W**) of the grain acts to pull it vertically to the bed, but the forward flow of the water exerts a force (**F**) to push the grain along parallel to the bed. The net result is for the grain to follow a path oblique to the bed, indicated by OD (*Figure 3.6*), which is in the direction

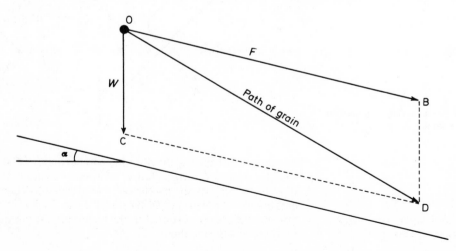

Figure 3.6 Forces acting on a grain in a flowing river

60

of the ***resultant*** of ***W*** and ***F***. The resultant direction and magnitude can similarly be found by completing a parallelogram of forces (OBDC) where OB and OC are proportional to ***F*** and ***W*** respectively, and CD and BD are drawn parallel to the opposite sides. We will see later on that the above example is a gross oversimplification of forces in flowing water, but it demonstrates the point of force-combination adequately.

3.1.5 Moments of forces

Sometimes forces do not cause simple forward motion of sediment grains but cause them to rotate about a point. This usually occurs when the grain is restrained at some pivot point; which would occur with a pebble resting on other pebbles on a fairly rough stream bed (***Figure 3.7***).

When a force tends to exert a rotation about a pivot it is said to exert a ***moment*** about that point. The moment of a force is equal to the product of the force and the perpendicular distance between the pivot and the line along which the force acts. Thus, if it is

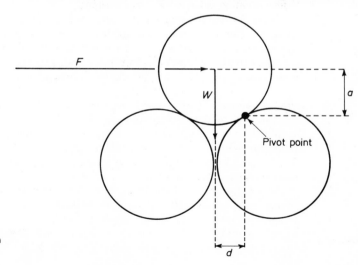

Figure 3.7 Moment of forces acting on a pebble resting on an uneven bed

assumed that the forward force of water flowing over the pebble (F) acts exactly through its centre, the moment due to F is equal to Fa, and that due to W (acting vertically through the centre of the grain) is Wd. That due to F tends to roll the grain along the bed but that due to W is tending to keep it in place. Clearly, when $Fa = Wd$ the grain is just on the point of moving and we can call this a limiting or *critical* force balance. The analysis of sediment transport processes often depends on identifying and quantifying critical static equilibria, which are indicative of the initiation of processes.

3.1.6 Forces of resistance

All slopes and the material upon them are under the influence of gravity, tending to pull them to the horizontal. Since much of the earth's land surface is *not* horizontal, it follows that there are forces which resist gravity and other forces tending to cause movement. So far these forces of resistance have been called reaction forces, since these are the ones which balance movement forces.

There are two main types of resistance to sediment transport, namely friction and cohesion. *Friction* is the force which tends to resist sliding of one object over another, for example, a mass of soil or a boulder sliding down a bedrock slope. Friction occurs wherever there is differential movement and is therefore present whenever one solid body slides past another, or when a fluid flows over a solid, or even over another layer of fluid. For the time being only solid to solid friction will be discussed, which is due to irregularities present on all surfaces no matter how smooth; and which tend to cause interlocking and binding. At the microscopic scale friction is a very complex mechanism but macroscopically it can be associated directly with the roughness of the surfaces involved. *Cohesion* is a force which chemically or physically 'glues' sediment grains together. It is discussed more fully in 3.2.4 since cohesion is often chemical in nature, but the fundamental difference between it and friction is mentioned below.

About the simplest case of frictional resistance in slope processes is that between a single boulder and a relatively smooth bedrock surface (*Figures 3.4* and *3.5*). Imagine first a boulder on a horizontal surface subjected to a horizontal force (F) which is steadily increased until the boulder begins to move (*Figure 3.4(b)*). Sliding will commence when the applied force (F_{crit}) is just equal to the *maximum* frictional resistance which can be developed ($R_{f\,crit}$). This is a critical static equilibrium case for sliding and marks the initiation of movement. It has been demonstrated experimentally that

$$F_{crit}/N = \text{constant} = \mu$$

where N is the normal reaction (in this case equal to the boulder's weight and therefore R_W) and μ is the coefficient of friction. Thus,

$$F_{crit} = R_{f\,crit} = \mu N$$

(where $R_{f\,crit}$ is the maximum frictional resistance) or frictional resistance is *proportional to the normal reaction*. Basic principles of friction can be shown by the simple experiments in 3.1.7.

Movement of material on slopes due to gravity is an important component of hillslope processes and can be illustrated by a boulder on an inclined surface. How steep does the slope have to be before the boulder will begin to slide under the influence of its own downslope weight component? The forces are shown in *Figure 3.5*. If the surface is tilted until the boulder just begins to move (again, the critical case for transport initiation) we know that $R_{f\,crit} = W \sin \alpha_{crit}$ ($W \sin \alpha_{crit}$ is the downslope component of the boulder's weight). The normal reaction in *Figure 3.5* is equal to $W \cos \alpha_{crit}$, since the boulder is on a slope of angle α_{crit}. Thus

$$\mu = F/N = \frac{W \sin \alpha_{crit}}{W \cos \alpha_{crit}}$$

when sliding begins, or

$$\mu = \tan \alpha_{crit} \qquad (\frac{\sin}{\cos} = \tan)$$

Hence the coefficient of friction is equal to the tangent of the slope angle at which sliding just begins. This critical slope angle, also a constant, is usually designated ϕ_μ, the *angle of plane sliding friction*. The experiments in 3.1.7 shows how this angle may be determined. Thus, one can write down the *strength* of the frictional contact between an object and a surface as:

$$F = N \tan \phi_\mu$$

since

$$\mu = F/N$$

and

$$\mu = \tan \phi_\mu$$

And since at sliding the frictional resistance $R_{f\ crit}$ is equal to the applied force (F_{crit}):

$$R_{f\ crit} = F_{crit} = N \tan \phi_\mu$$

Thus, frictional resistance is resistance to movement which is proportional to the normal reaction (N). Cohesion, by contrast, is strength which is *independent of normal reaction*. It acts like a glue and is *additive* to strength. The mechanisms of cohesive strength are discussed in 3.2.4.

3.1.7 Experiments on the principles of frictional resistance

These experiments are designed to show the fundamentals of frictional resistance to movement.

3.1.7(a)

The coefficient of friction is constant for any given materials and frictional resistance is proportional to the normal reaction. To demonstrate this, weigh a flat sided wooden block (or a housebrick) and call this weight W. Place it on a horizontal surface and attach a spring balance. Pull the balance slowly and steadily until the block just begins to move and record the reading on the balance as F. Repeat this a number of times to obtain a mean value for F. The coefficient of plane sliding friction (μ) for that block on that surface can be calculated from $\mu = F/N$, where N is the normal reaction equal to W.

Repeat the experiment three or four times with known weights added to the block each time (thus increasing W). The calculated values of μ at each repeat should be close to each other, though there will be some experimental error. When the block just starts to slide frictional resistance (R_f) just equals the applied force (F) so that:

$$R_f = F = \mu N$$

and since μ is a constant, $R_f \propto N$. This means that frictional resistance to movement is proportional to the normal reaction.

3.1.7(b)
The results from 3.1.7(a) can be used to calculate the angle of sliding friction. If we were to incline the board on which the block stands instead of pulling the block with a spring balance, at what angle of inclination would the block begin to slide?

Plot the results of 3.1.7(a) on the graph and the angle of sliding friction (the angle at which the block begins to slide when tilted) is the angle between the best-fit line through the experimental points and the x-axis of the graph. (This can be measured with a protractor or calculated as the angle whose tangent is F/N, i.e. as the arctangent of μ.)

3.1.7(c)
The same experimental design can be used to show that friction is independent of the area of contact. Repeat 3.1.7(a), but placing the block or brick on different sides which have different areas. The reading on the spring balance (F) at the point of movement should be the same provided that the different sides have the same roughness.

3.1.7(d)
The experiment 3.1.7(c) would not work if different sides of the block had different roughnesses. This is because μ, the coefficient of friction, while being independent of

area of contact, will vary with surface roughness. Prepare three wooden boards and three wooden blocks one of each being polished, planed and rough (e.g. covered with sandpaper). Repeat 3.1.7(c) for each block on each surface and compile a table of μ values as below. How do the angles of sliding friction vary?

CALCULATED MEAN ANGLES OF SLIDING FRICTION

		BOARDS		
		polished	planed	rough
BLOCKS	polished			
	planed			
	rough			

3.2 CHEMICAL ASPECTS OF HILLSLOPE PROCESSES

Most hillslopes are not composed of the kinds of chemically inactive solids which formed the basis of discussion in the previous section. They are, for the most part, mantled with chemically active soils and, in order to be able to understand the behaviour of hillslopes better, we must consider some of their chemical properties and the chemical processes occurring on them.

Cohesion between particles may be chemical in nature and is additive to the frictional forces resisting movement considered in 3.1.6. Water moving downslope under the influence of gravity has the power to take into solution part of the fabric of the materials through and over which it flows and this is an important process of erosion on hillslopes. Solution operates selectively on certain constituents and therefore alters the chemical composition of bedrock and soil. This alteration is known as chemical weathering and since we are only able to discuss this very briefly here you may find it helpful at this stage to refer to an introductory text on weathering. For our purposes the most important residual products of weathering are the clay minerals because of their influence on hillslope stability and because they are the sites of much chemical activity.

3.2.1 Atomic structure and bonding

All matter is made up of atoms. An atom consists of a central nucleus surrounded by concentric electron orbits, arranged so that the negatively charged electrons balance the positive charges on the protons in the nucleus. A standard number of electrons occupy each orbit and these orbits are usually called *shells*. The first electron shell, when full, will contain only two electrons, while all subsequent shells will contain up to eight electrons. The chemical stability of an element is determined by whether or not the outer electron shell is full. Thus hydrogen, which has only one proton in the nucleus, has only one electron. Since it requires another electron to complete that first shell it enters readily into chemical reactions. Helium on the other hand does not, since it has two protons and therefore a complete first shell of two electrons. Similarly we can compare a reactive element like oxygen with a stable element like neon and see that the second shell in neon has eight electrons while that of oxygen has only six (*Figure 3.8*). This is a rather simplistic view of atomic structure but useful for our purposes.

Atoms tend to combine together so as to complete their outer electron shells, either by sharing electrons or by transferring them. When the electrons are shared a *covalent* bond forms as is the case in the water molecule (*Figure 3.9*), where one oxygen shares an electron with each of two hydrogen molecules and the hydrogens share their electrons with the oxygen. Thus the outer shells of the oxygen and the hydrogen molecules are completed.

The transfer of electrons produces *ionic* bonds. For example, sodium has only one electron in its outer shell while chlorine only needs one to complete its outer shell (*Figure 3.9*). When the electron is transferred from the sodium to the chlorine it leaves a net positive charge on the sodium and causes a net negative charge on the chlorine. These two charged particles, or *ions* are then held together by electrostatic attraction.

Positively charged particles formed in this way are called *cations* (e.g. sodium, potassium, calcium, magnesium) while the negatively charged particles are *anions* (e.g.

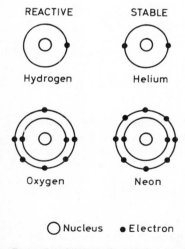

REACTIVE — Hydrogen

STABLE — Helium

Oxygen

Neon

○ Nucleus ● Electron

Figure 3.8 Models of reactive and stable elements

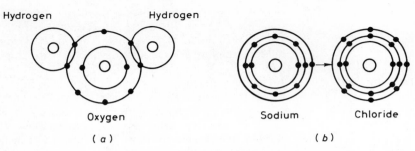

Hydrogen   Hydrogen

Oxygen

(a)

Sodium   Chloride

(b)

Figure 3.9 Interatomic bonds as in water and sodium chloride

oxygen, chloride, hydroxyl). The number of electrons which are either donated by cations or accepted by anions determine the **valency** of the ion. For example sodium (Na^+), hydrogen (H^+), and chloride (Cl^-) all have a valency of one while oxygen (O^{--}), calcium (Ca^{++}) and magnesium (Mg^{++}) all have valencies of two.

Bonds of both types are common in soil forming minerals. Ionic bonds and in some cases covalent bonds (e.g. water) produce dipolar molecules which have one end positively charged and the other end negatively charged.

Covalent and ionic bonds are relatively strong primary bonds but weaker secondary bonds also form between particles. These secondary bonds are very important sources of attraction between small particles and between liquids and solids.

When hydrogen is the positive end of a dipole, as in the water dipole, a relatively strong secondary bond, called a **hydrogen bond**, can be formed with certain electronegative atoms, particularly oxygen. The strength of the hydrogen bond is due to the very small size of the hydrogen ion.

Other dipole bonds also exist and are commonly called **Van der Waals bonds**. While Van der Waals bonds are much weaker than hydrogen bonds their strength decreases less rapidly with distance than do other types of bonds.

3.2.2 Clay mineral structure

Clay minerals are formed by the chemical weathering of silicate minerals. The primary silicate minerals are chemically unstable in the weathering environment and either decompose completely, and more stable minerals are formed from the decomposition products, or undergo physical and chemical alteration to form new minerals. In both cases some of the constituents of the original minerals are lost in solution in drainage waters.

Clay minerals or colloids are very small crystals, usually between $0.2\,\mu m$ and $2\,\mu m$, and are made up of layers of silica (SiO_2) and alumina ($Al(OH)_3$). The silica sheet (*Figure 3.10*) is made up of silicon atoms in tetrahedral (four-sided) coordination with oxygen. Each silicon has a valency of four and each oxygen a valency of two, so that

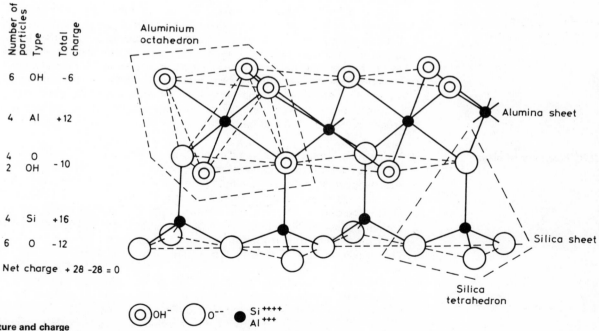

Number of particles	Type	Total charge
6	OH	-6
4	Al	+12
4	O	-10
2	OH	
4	Si	+16
6	O	-12

Net charge + 28 -28 = 0

$\bigodot$ OH^- $\bigcirc$ O^{--} $\bullet$ Si^{++++} Al^{+++}

Figure 3.10 Structure and charge distribution of kaolinite

69

the charges are balanced. The alumina sheet (*Figure 3.10*) is made up of aluminiums (Al^{+++}) in octahedral (eight-sided) coordination with oxygen (O^{--}) or hydroxyl (OH^-). The silica and alumina sheets are bound together by shared oxygen atoms.

In the ideal situation shown in *Figure 3.10* all the charges are balanced. In practice, atoms of similar size but lower valency can replace the silicon and aluminium leaving a net negative charge on the crystal surface. Aluminium is similar in size to silicon but has one less positive charge. Aluminium is commonly substituted for silicon in the silica sheet. Ferric iron (Fe^{+++}), magnesium (Mg^{++}), zinc (Zn^{++}) and ferrous iron (Fe^{++}) are all of similar size to aluminium and may be substituted in the alumina sheet. The net negative charge produced by this substitution attracts or *adsorbs* cations and the positive ends of dipoles to the crystal surface. These negative charges are called cation exchange sites and give to the clay minerals a *cation exchange capacity* which varies between different clay mineral types. In effect this means that clay minerals can adsorb cations because they are attracted to the negatively charged sites. In time these cations are often displaced by hydrogen ions from water draining through the soil. We will return to this subject when processes of solution are discussed in Chapter 5.3.

3.2.3 Plasticity of clays

You will have noticed that wet clay has a sticky feel to it and can be moulded. This handling consistency is termed *plasticity*. Plasticity is best defined by reference to the *Atterberg limits* of clays. The liquid limit is the minimum moisture content (expressed as a percentage of dry weight) at which the clay flows under its own weight. The *plastic limit* is the minimum moisture content at which the clay can still be moulded. At moisture contents above the liquid limit the clay behaves as a viscous liquid, between the plastic limit and the liquid limit the clay is plastic and can be moulded, and below the plastic limit it behaves as a brittle solid. The plasticity of a clay is the range of moisture content over which it remains plastic.

There are three main types of clay minerals, kaolinite, montmorillonite and illite. The main properties of each are discussed below and summarised in *Table 3.1.*

Kaolinite is made up of one silica and one alumina layer. The layers of kaolinite are rather rigidly bound together by hydrogen bonds and Van der Waals bonds. The effective surface is therefore restricted to the outer surface of the units. This, together with the fact that there is little substitution in the sheets, gives it low plasticity, low cation exchange and low swelling capacity.

Montmorillonite is composed of one alumina layer between two silica layers. The layers are bound together by Van der Waals bonds and adsorbed cations which are weak enough to allow water to enter. Extensive substitution occurs to give the crystals a high net negative charge. Consequently cation exchange capacity, plasticity and swelling capacity are high.

TABLE 3.1 COMPARATIVE PROPERTIES OF THREE MAJOR TYPES OF CLAY MINERALS. (AFTER BRADY, 1974)

Property	Type of clay		
	Montmorillonite	Illite	Kaolinite
Size (μm)	0.01–1.0	0.1–2.0	0.1–5.0
Shape	Irregular flakes	Irregular flakes	Irregular flakes
Internal surface	Very high	Medium	None
External surface	High	Medium	Low
Plasticity	High	Medium	Low
Swelling capacity	High	Medium	Low
Cation exchange capacity	High	Medium	Low

Illite or 'hydrous mica' has a structure similar to that of montmorillonite. In illite a significant proportion of the silicon atoms have been replaced by aluminium giving the silica layers a high net negative charge. Potassium ions are attracted to these sites and serve to bind the silica layers together thus reducing their swelling capacity. In terms of swelling capacity, cation exchange capacity and plasticity, illite is intermediate between kaolinite and montmorillonite.

As suggested earlier, chemical weathering serves two main functions which have important consequences for hillslopes. It alters the primary silicate minerals to form clays which play a significant role in determining the physical stability of slopes, and ions released by weathering are carried away in solution thus contributing to hillslope erosion.

3.2.4 Cohesion

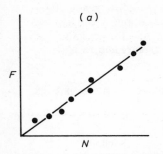

Figure 3.11a Experimental results from 3.1.7(a)

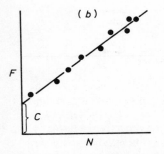

Figure 3.11b For the same experiment where cohesion is present

If we were to take a set of results from experiment 3.1.7(a) and plot them on a graph as in *Figure 3.11* we can see that a line joining the experimental points will pass through, or at least very close to, the origin (i.e. the point where $F = 0$ and $N = 0$). We know from 3.1.6 that the equation of the line can be written as:

$$F_{crit} = N \tan \phi_\mu$$

Cohesion adds to the strength of a contact between a block and a surface in the same way as a glue. With some cohesion present between the block and the surface the results should look like *Figure 3.11(b)*. Notice that this time the line does not pass through the origin but makes an intercept on the y-axis, C, which is due to cohesion. The equation of the new line is:

$$F_{crit} = C + N \tan \phi_\mu$$

which is known as the Coulomb equation. Notice that the cohesion simply adds to the strength.

In the materials forming hillslopes, cohesion can be due to capillary cohesion, or chemical bonds of the type we discussed earlier. Cohesion derived from capillary water was discussed in 3.1.1.

Cohesion between particles may also be caused by *cementation* where the chemical bonds are of the primary type and therefore very strong. Common cementing materials are carbonates, silica, alumina, iron oxide and organic compounds. The importance of this type of cohesion is demonstrated by the experiment in 3.2.5. Cementation may also develop from compaction when the material is subject to high pressure such as would be found at depth below the ground surface. Cohesion can develop in clays which have been buried and the overburden stripped off by subsequent erosion. Such clays are said to be *overconsolidated* and have much greater cohesive strengths than similar materials which have not been compacted.

As we have seen, cohesion can also result from attractions between very small particles. The most important of these is apparently the Van der Waals bond although this type of cohesion is not yet fully understood. It is clear from our discussion of clay minerals that since the surfaces of clay colloids are electrically charged the forces of attraction between them will be much greater than the attractions between similar sized but inert particles, such as very small grains of quartz. The bonds between the clay colloids are known as 'long range bonds' and will remain active when the colloids are moved relative to each other, hence the 'plasticity' of clays. The bonds between small inert particles lose all their strength when relatively small displacements occur and are therefore termed 'short range bonds'. The difference in behaviour of the two types of bonds is illustrated in *Figure 3.12*.

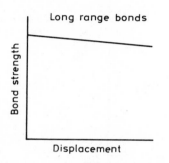 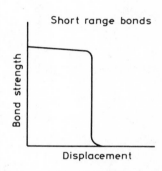

Figure 3.12 Behaviour of long and short range bonds during displacement

The relative importance of different types of cohesion in contributing to strength is shown in *Figure 3.13*. Notice that interparticle attractive forces are significant only for small grain sizes. While these forces are small relative to cementation they are still very important components of the total strength of unconsolidated material on hillslopes, the behaviour of which is usually dominated by the clay mineral fraction.

3.2.5 Experiment to demonstrate cohesion

Cohesion between particles adds to their strength and enables slopes formed by those particles to stand at higher angles. Take two 250 ml glass beakers and fill them with *clean*, coarse sand (wash the sand if necessary). Saturate the sand in one beaker with distilled water. Make up a supersaturated solution of salt by heating some water and adding salt

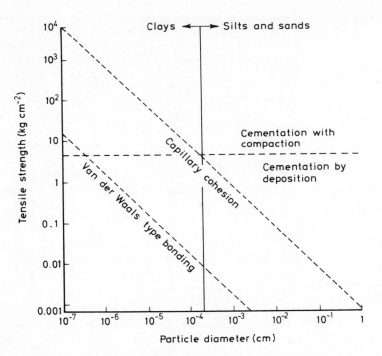

Figure 3.13 Contribution of various bond types to soil strength (After Mitchell, 1976)

until no more will dissolve. Saturate the sand in the other beaker with this salt solution. Turn 'sand castles' out of the two beakers and leave to dry on a tray. (The drying can be speeded up by placing them in an oven, if necessary.)

When the moisture has evaporated the sand castle made with distilled water will collapse into a conical pile while that made with the salt water will not, since the individual grains are glued together with salt. This experiment also demonstrates the importance of capillary water as a source of cohesion in the damp sandcastles.

3.2.6 Chemical denudation

Chemical denudation is caused by the removal in solution of soluble mineral components. The effectiveness of this process is controlled by many factors: the equilibrium solubility of the minerals present; the amount of water passing through the system; the physico-chemical nature of that water; its temperature; and finally length of time in contact with the minerals to be dissolved, which is governed by the rate of flow. The removal of material in solution is necessary if weathering is to continue unimpeded, otherwise the weathering products accumulate around the primary minerals and inhibit the rate of weathering. This explains many of the differences in weathering rates and products between arid and humid environments.

The solubility of individual compounds varies with changes in the physicochemical state of the water and the state of the compound itself. One of the most important determinants of solubility is pH which is *a measure of the concentration of hydrogen ions*.

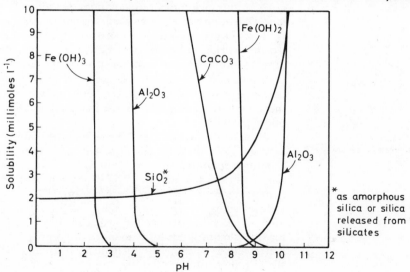

Figure 3.14 Solubility in relation to pH for some components released by chemical weathering (After Loughnan, 1969)

pH is expressed as the logarithm of the reciprocal of the hydrogen ion concentration so that pH decreases as the hydrogen ion concentration increases. a pH of 7 is neutral while values above 7 are alkaline and below 7 acid. Most natural waters lie in the pH range 4–9.

The relationship between pH and some of the major components released during weathering is shown in *Figure 3.14*. It can be seen from this diagram that certain important mineral constituents, particularly ferric iron and aluminium oxide, are completely insoluble in the pH range of natural waters and so are not removed in solution. Amorphous silica has a low but relatively constant solubility in the pH range 4–9 but the solubility of silica as quartz is only about one third of this level. The solubility of calcium carbonate and ferrous iron is pH-dependent. Sodium and potassium (not shown in *Figure 3.14*) are very soluble and readily removed by water draining through the soil.

The solubility of some elements, particularly iron, is affected by the redox potential or E_h of the environment. *Redox potential refers to the transfer of electrons to and from ions. Oxidation* occurs when electrons are removed and *reduction* when they are added. Under oxidising conditions iron exists as a trivalent ion (Fe^{+++}) and is completely insoluble above pH 3. In the absence of oxygen, i.e. reducing conditions, iron exists as a divalent ion (Fe^{++}) and is soluble below pH 9 (*see Figure 3.14*).

It follows of course that changes in the environment can cause both increases or decreases in the amount of a given compound in solution. If water in a reducing environment, with iron in solution, passes into an oxidising environment, iron will be precipitated out of solution.

When a mineral dissolves in water and the material dissolved from the mineral has the same composition as the solid phase it is said to dissolve congruently or to form a *congruent solution*. An example of this is the solution of calcium carbonate in water. In pure water the amount of $CaCO_3$ which will dissolve is limited by the availability of hydrogen ions to form stable bicarbonate (HCO_3^-) ions from the carbonate. When CO_2 is dissolved in the water much more $CaCO_3$ will go into solution because two bicarbonate ions are needed to balance the two positive charges on each calcium ion. In soils, carbon dioxide is more abundant than in the atmosphere because of the activity of soil organic matter and so calcium carbonate dissolves readily. When water, saturated with $CaCO_3$ in the soil, flows out into the free atmosphere calcium carbonate may be precipitated as the solution comes into equilibrium with the lower CO_2 content (usually referred to as the partial pressure of CO_2). This illustrates the effects of environmental conditions on solution and precipitation. A good example is the formation of stalactites and stalagmites in limestone caves.

When a mineral dissolves in water such that the composition of the dissolved phase is different to that of the solid phase it forms an ***incongruent solution***. This is more correctly termed a reaction between the mineral and the water. When water comes into contact with a clay mineral, say kaolinite, the silicon in the mineral will go into solution far more readily than the aluminium so that we are left with some of the silica in solution and an altered solid phase left behind (in this case an aluminium hydroxide). In this way minerals change their structure and composition while releasing some part of their original fabric into solution.

On hillslopes both congruent and incongruent solution occurs and water moving down a hillslope changes its composition and encounters varying environments on its way with which it attempts to come into equilibrium. The extent to which this equilibrium is attained depends on the rate of flow of the water since the chemical reactions take time. We will return to this point in Chapter 5.3.

CHAPTER 4   MASS MOVEMENT PROCESSES

4.1 DEFINING MASS MOVEMENT

Mass-movement is a term applied to a great variety of phenomena which displace bodies of material. Strictly speaking it includes subsidence of the ground surface (as above mine galleries) and movements which occur on the sea bed, but here we will be discussing only those processes which occur on land-surface hillslopes. The important distinction between mass movement and other transport processes on slopes is that in mass movement no transporting medium such as ice, air or water, is involved. *Mass movement*, sometimes called *mass-wasting*, is the *movement of rock and soil debris downslope under the influence of gravity and without the aid of flowing water, wind or glacier ice.*

In Chapter 1 (*Figure 1.3*) we saw that in the erosional development of a river valley the greater part of all the material eroded to form the valley must be moved to the river by slope processes. The same is true, though to a lesser extent, of valleys containing glaciers. The function of mass movement is to transport rock and debris downslope to the valley bottom where it can be picked up by a river or a glacier. Mass movement is also common along coasts where it moves material into a position where it can be transported away by the sea. We also saw in Chapter 1.2 that mass movement is one of the main concerns of engineers in dealing with both natural and artificial slopes.

4.1.1 Classification of mass movement

A number of different classifications of mass movement have been proposed and very many terms have been coined to describe the types of movement which occur. A simplified classification is shown in *Figure 4.1*. The types shown in the classification may move in one (or some combination) of four different ways. *Sliding* is movement along a shear plane where all the moving material travels at the same velocity. It occurs in hard rocks as slab and wedge failures and in soft rocks as rotational slips, translational slides and mudslides. The term mudslide, used here, is replacing the hitherto more usual term 'mudflow', because the movement is now recognised to be commonly sliding rather than true flow.

Figure 4.1 Classification of mass movement

A: FAILURES IN HARD ROCK

A I Controlled entirely by discontinuities (e.g., joints and bedding planes)

AI(i) Slab failure

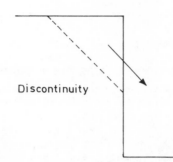

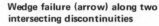

Slab failures (arrow) along
steeply dipping discontinuities

AI(ii) Wedge failure

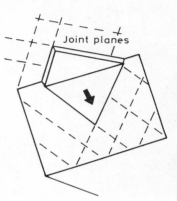

Joint planes

Wedge failure (arrow) along two
intersecting discontinuities

AI(iii) Toppling failure

Toppling failure (left, arrow) due to vertical stress-relief joints

AII Not controlled solely by discontinuities

Large scale rock avalanche in alternating lavas and ashes, Iceland (below)
Note slip scar (arrow, left) and avalanche debris (arrow, right)

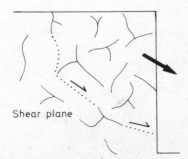

Shear plane

Rock avalanche

AIII Controlled by rock disintegration

Rockfall

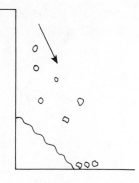

B: MOVEMENT IN CLAY AND WEATHERED REGOLITH

BI Rapid failures

BI(i) Rotational slip

Rotational slip in Mesozoic strata, Dorset. Note degraded shear plane (arrow, left) and rotated block (tilted building) (arrow, right)

BI(ii) Translational slide

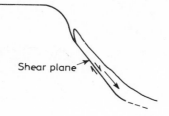

Shear plane

Translational slide in glacial lake clays, Eire
(arrow marks shear plane)

BI(iii) Loose particulate avalanche

Loose particulate avalanche slope accumulating by
degradation of a fluvioglacial sand cliff, Eire

82

BII Rapid failures with high water content

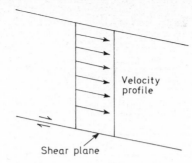

Velocity profile

Shear plane

BII(i) Mudslide

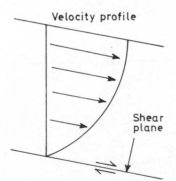

Velocity profile

Shear plane

BII(ii) Debris flow

Debris flow gully and trail, Black Mountains

BIII Slow movements

**BIII(i) Soil creep: (a) seasonal;
(b) continuous; (c) random.
(ii) Solifluction**

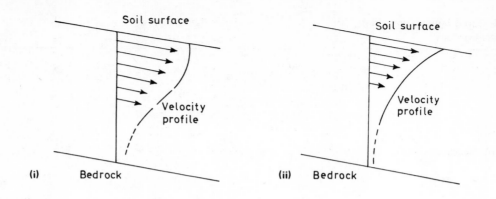

(i)

(ii)

Flow is movement in which velocity increases away from the shear plane. The distinction between flowing and sliding is illustrated in the explanatory diagrams to category B II in *Figure 4.1*.

When failures take place on very steep slopes, such as cliffs, the material *falls* rather than slides. Most of the failures in hard rocks illustrated in *Figure 4.1* are usually falls, because they generally occur on very steep slopes. They occur also in soft rocks as, for example, soil falls or the undercut banks of streams.

Creep is the very slow downslope movement of soil and regolith. Movement is fastest at or near the surface and diminishes with depth so that no shear plane separates the moving and stationary material.

4.1.2 Rates of movement

Some estimates of the rates of movement of these processes are shown in *Figure 4.2*. The rapid failures range from a few centimetres per day for some debris flows to more than 300 kilometres per hour for some failures in hard rock. As *Figure 4.2* shows, debris

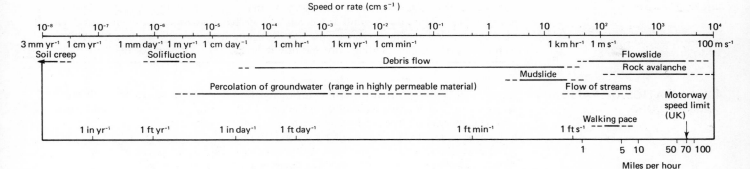

Figure 4.2 Ranges of rates of mass movement and water flow

flows have a very wide range of rates, partly controlled by slope angle, but also usually closely related to water content. Mudslides have a much narrower range of speeds but for material to be classified as mud it must have a high water content, and hence the high rates of movement. The slow processes are mainly less than a few millimetres per day and more often only a few millimetres per year. However, this does not mean that the slow processes are less important than the rapid ones. Slope failures occur only under a restricted range of conditions but attract a great deal of attention because they are usually quite spectacular and may present serious hazards to human activity. The slow processes such as soil creep are not spectacular or hazardous but, because they occur more or less continuously on nearly all slopes, are very important in landscape development.

The classification of mass movement given in *Figure 4.1* does not adequately represent the processes which occur in the real world. Mass movements do not occur as discrete types but rather are ranged along a continuum. Any individual occurrence of mass movement may involve a combination of two or more types or may be somewhere between the specific types mentioned here. For example, rotational slips on clay slopes often grade into mudslides at the base of the slope with the blocks of failed clay feeding the mudslide. Similarly, the distinction between mass movement as defined here and transport by water and ice is quite arbitrary. In fact, a continuum exists between these two sets of processes. Some streams become so highly charged with solids that they begin to behave like debris flows, while some debris flows have such high water contents that they begin to behave like streams. This classification, like all classifications, should be seen as an aid to understanding only and not as an accurate portrayal of reality.

4.2 INITIATION OF RAPID MASS FAILURE

It should be quite apparent from the previous section that rapid mass movements are very diverse in character, depending on the type of soil or rock involved, its water content, and the scale of the process. The latter part of this section explores some of the influences of soil and rock characteristics on rapid mass failure style. But for the time being we will dispense with the complexity of landslide style and consider the very simple underlying mechanical principles governing landslide operation.

4.2.1 Mechanics of landsliding

Rapid mass movements or landslides involve rapid lateral displacement of a large mass of soil or rock along a well defined planar break within the ground. The plane, called a *shear plane* because the soil is *sheared* or split along it, is usually a recognisable rupture surface and practically all the movement takes place along it. Consequently, the landslide mass itself often retains much of its original shape, since it moves as a more or less coherent block of material with little internal disturbance. Naturally, exceptions occur to this, especially when the slide takes place on a steep slope or cliff and tends to break up as it falls. Rates of operation of so-called rapid mass movements are very variable, a point already made in Chapter 4.1, and depend very much on the type of material and steepness of slope. But the important point to note is that they always operate fast enough to cause visible disruption of the ground surface. They are therefore one of the most visually obvious processes operating on slopes.

Forces in landslides
The underlying controlling mechanics of landsliding are very simple indeed. Since movement takes place on a well defined plane it follows that sliding will occur when *the forces tending to cause movement are greater than those resisting it*. How and when the process begins is considered in 4.2.5. For the time being we must consider the origin of the forces

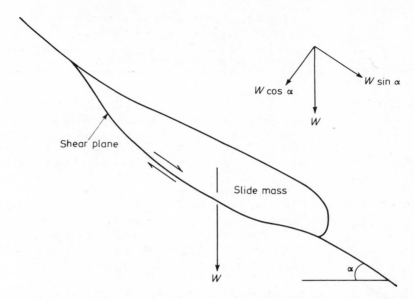

Figure 4.3 A simple landslide

themselves, which will involve taking up some points which were first introduced in 3.1.3, 3.1.6 and 3.1.7.

The force tending to **cause** movement of a landslide is, of course, derived from gravity. We can think of a landslide as a body resting on a slope (***Figure 4.3***), and the force tending to drive it downslope is the component of its own weight, acting parallel to the slope. In the case illustrated this force is equal to $W \sin \alpha$.

The forces tending to **resist** movement of the landslide depend on the strength of the material involved. Before the slide starts, and therefore before the shearplane develops, we must consider the ***intact strength*** of the soil or rock. Intact strength, the strength of the soil or rock prior to any disturbance by sliding, is obviously much greater than the strength across the shear plane once it has formed, because the shearplane is a break in the material. The lower strength along the shearplane after the slide has moved some distance is called the *residual strength*. Following from 3.1.6, strength is made up of frictional resistance and cohesion. A very simple law called *Coulomb's law* describes the strength of a soil, namely:

$$S = c + \sigma \tan \phi$$

where S is strength, c is the cohesion, σ is the normal stress in the ground and ϕ is the angle of internal shearing resistance of the soil or rock. We can appreciate this law by referring to 3.1.6 and *Figure 3.5*. It was stated in 3.1.6 that the frictional resistance (R_f) to the movement of a block down a slope (*Figure 3.5*) is equal to:

$$R_{f\,\text{crit}} = N \tan \phi_\mu$$

where N is normal reaction at the slide surface and ϕ_μ is the angle of plane sliding friction for the block on the slope. It was also said that cohesion is additive to the strength of a material, rather like a glue. Hence the total resistance (R) to the sliding of the block in *Figure 3.5* were it to possess some cohesion, would be:

$$R = C + N \tan \phi_\mu$$

where C is the cohesive force between the block and the surface (Chapter 3.2.4). Coulomb's Law is exactly analogous to this equation, except for two small modifications. First, instead of the equation being written in terms of *forces* as in Chapter 3, we write it in terms of *stresses*. Stress is simply defined as force per unit area (force divided by the area over which it operates). Thus we represent the cohesive *stress* as c and the normal *stress* across the shearplane as σ (instead of C and N respectively). Secondly because a landslide slips over a plane developing between soil grains or within a rock, and not as a rigid block over a smooth surface, we replace ϕ_μ with ϕ, the *angle of internal shearing resistance*. The angle of internal shearing resistance for a loose, dry pile of cohesionless soil (sand or gravel) is approximately equal to the maximum angle of slope it would stand at before beginning to slide. This maximum angle of slope is sometimes called the **upper angle of repose**. The **lower angle of repose** is the angle at which the sliding mass comes to rest. The reason for the difference between ϕ and ϕ_μ is that a shear plane in soil must pass between the grains and is not perfectly flat (**Figure 4.4**). Thus before movement can take place, interlocking of the grains must be overcome. (The experiment in 4.2.6 shows you how to measure critical stable angle for dry soils and is intended to follow on from the simple friction experiments described in 3.1.7.)

We now know the forces causing and resisting movement of a landslide. **Just at the point when sliding commences they are equal.** Therefore we can write:

$$T_f = S = c + \sigma \tan \phi$$

where T_f is the **shear stress** (stress tending to cause lateral movement on the shearplane) at failure. Simply then, **shear stress at failure is equal to shear strength.**

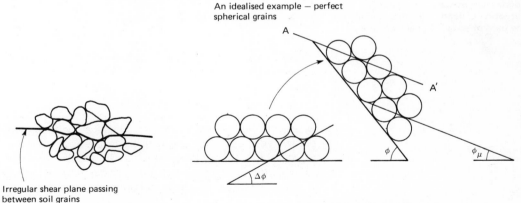

An idealised example — perfect spherical grains

Irregular shear plane passing between soil grains

Figure 4.4 Shearplane development within a soil

Sliding will not begin until plane A — A' is inclined at ϕ_μ
$\therefore \phi = \phi_\mu + \Delta\phi$

89

The role of water

One other important factor needs to be considered in the stability of a landslide, namely the presence of water in the soil or rock. Water in the pore-space between soil grains (*Figure 3.1*) or in the cracks in rocks, exerts a pressure on the surrounding material. Provided the soil is saturated this is a positive pressure, which at any point is proportional to the head of water above that point. A simple case is illustrated in *Figure 4.5*. This

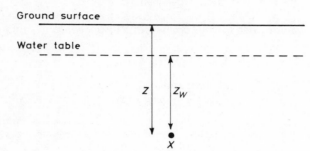

Figure 4.5 A simple illustration of porewater pressure. Porewater pressure, at depth Z is proportional to head of water (Z_w) above point X

pressure reduces the normal stress between soil grains by pushing them apart, and so Coulomb's Law becomes:

$$T_f = S = c + (\sigma - u) \tan \phi$$

where u is the porewater pressure. Thus the frictional strength is reduced due to the porewater pressure, making sliding possible on lower slope angles. It is very important to recognise that the key effect of water in landsliding is this pressure effect, and is not due to lubrication or increase of the soil weight, as is often popularly stated.

To summarise the simple mechanics of landsliding one can say that:

(1)   *The forces tending to cause a slide* are due to the downslope component of the weight of the slide mass.

(2)   **The forces tending to resist movement** are dependent on the strength of the material, which consists of friction and cohesion.

(3)   Water in soil pores (or rock cracks) exerts a pressure on the surrounding material, reducing the frictional strength and making sliding easier.

Landsliding in cohesionless soils is only governed by the slope angle, in that there is a critical angle at which sliding begins. For dry soils it is approximately equal to ϕ. When the soil is filled with water, however, the critical stable angle is less than ϕ, due to the porewater pressure which reduces frictional strength. Since slope angle is the only geometric control on sliding, landslides in cohesionless materials tend to be long and shallow, with the shearplane nearly parallel to the ground surface (*Figure 4.6*(a)). In cohesive materials, an extra component of strength is added to friction. Thus the **height** of the slope as well as the angle becomes important. Indeed, a cohesive material can form a vertical cliff for example, but there will be a critical height of cliff which is unstable. Landslides in cohesive materials tend to have deep seated shear planes which are often curved (*Figure 4.6*(b)). These are sometimes called **rotational** slips.

All landslides are broadly similar in the way they operate, in that they conform to the simple mechanical principles discussed above, but the style of landsliding and its influence

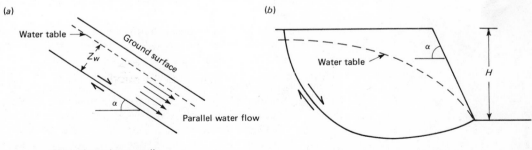

(a)

Water table

Z_w

Ground surface

α

Parallel water flow

Planar shear plane, usually near to junction between bedrock and soil.

(b)

Water table

α

H

Curved shear plane (rotational slip), often deep seated.

Figure 4.6 Typical landslide form in cohesionless and cohesive soils

on slope form (Chapter 6.2) depends on the type of material in which it occurs. The remainder of this section is therefore taken up with a discussion of the differences in landslide character between three broad groups of material:

hard, coherent rocks,
clays,
granular soils.

4.2.2 Failures in hard rocks

Hard rocks, including igneous rocks such as granite and basalt and sedimentary rocks such as sandstone and limestone, have very high cohesive strength. The cohesion in igneous rocks is due to the constituent minerals fusing together as they cool down, whereas in sedimentary rocks it is often due to a secondary mineral cement which fixes the grains tightly together. In fact, cohesion is so high in intact, unweathered, hard rocks that frictional strength is almost irrelevant by comparison. It is possible to calculate the maximum stable height of vertical cliffs composed of hard rocks and the result is always staggeringly high. The strongest rocks, for example granite and quartzite, should be able to stand as quite stable cliffs between 4 and 10 kilometres high! However, landslides frequently occur on slopes composed of these rocks only a few metres high, and so the intact strength of the rock cannot be a very good indicator of the real strength of rocks in the field.

The answer to this paradox is quite simple. All rocks are cut by *discontinuities*; breaks or cracks which weaken them considerably. These may be bedding planes, cleavage, joints or faults and any of them may be responsible for reducing strength of rocks very considerably. In many cases all cohesive strength is lost across a discontinuity (that is, one side is totally separated from the other) and only frictional resistance exists between the rock masses. It is therefore hardly surprising that most landslides in rockslopes are controlled fairly closely by the discontinuities which are present. The annotated photographs and diagrams in *Figure 4.1* show the main styles of rock failure.

At the smallest scale, failures of one or two joint-bounded blocks loosened by weathering are called *rockfalls*. This process will occur on any rock face as weathering weakens the rock along discontinuities, and it does not indicate that the whole cliff is unstable.

Medium-scale failures in hard rocks are usually totally controlled by the orientation and spacing of the discontinuities with respect to the face. Three typical failure styles are shown in *Figure 4.1*: *slab failure*, where sliding occurs on discontinuities dipping into the face; *wedge failure*, where two sets of discontinuities intersect to define wedges of rock dipping towards the face; and *toppling failure*, where undercut or unsupported blocks can fall out of the face or topple downslope. In each case, the discontinuity pattern controls the style of failure and the failures are entirely along discontinuities.

Large scale rock-slides or *rock avalanches* tend to have deep shearplanes which partly follow discontinuities and partly cut through intact rock. The character of the discontinuity surfaces is therefore always important in determining whether a slide will take place on a rock slope. Their roughness determines frictional resistance and the presence of mineral infills may affect cohesion across the planes. One should also remember that water pressures, which are often very considerable, can exist in discontinuities when they become water-filled, and will obviously affect strength. Some experiments are included in 4.2.7 to show you how to measure frictional strength of rock surfaces and to demonstrate how orientation of discontinuities controls the style of failure on rock slopes.

4.2.3 Failures in clays

Clays, such as Lias Clay or London Clay, also possess significant cohesive strength (Chapter 3.2.4). But clay cohesion, while being important, is much less than cohesion in hard rocks, and therefore frictional strength is always significant as well. Cohesion is really only present in clays which have been *overconsolidated*. An *overconsolidated clay* is one which has been subjected to a higher level of overburden pressure than at present. Thus it was once covered by more strata and has been re-exposed by erosion. The effect of overconsolidation is to expel water from the clay, force the grains closer together, and to increase the forces of attraction between them (*Figure 4.7*). In contrast, a normally consolidated clay has never been subject to higher overburden pressures than at present and cohesion is negligible (*Figure 4.7*).

Landslides in unweathered, overconsolidated clays are usually deep-seated rotational failures. Coastal cliffs in the Lias Clay of Dorset and the London Clay of the North Kent coast show typical large scale rotational slides in intact clay and illustrate the process very well (*Figure 4.1*). Normally-consolidated clays, on the other hand, usually fail in shallow

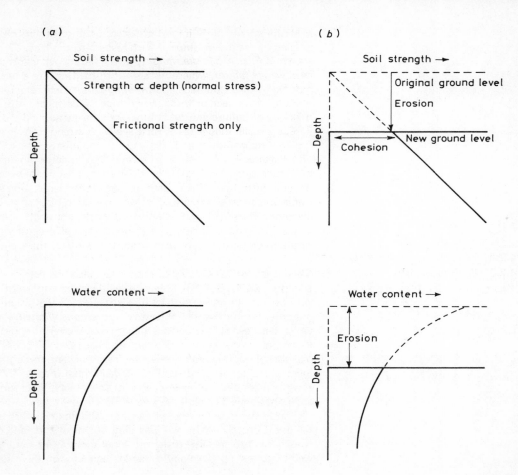

Figure 4.7 The strength and water content of (a) normally consolidated and (b) overconsolidated clays

94

landslides with the shearplane parallel to the ground surface, because they possess no cohesion. *Figure 4.1* shows such a slide in glacial lake clays.

Although clays are generally less affected by discontinuities than harder rocks, the presence of such features cannot be entirely ruled out. London Clay and Lias Clay, for example, possess many small cracks and fissures which form an interconnecting network throughout the material. Sometimes clays even possess well-defined joints which influence the style of landslides, as in hard rocks. Generally the effect of discontinuities in clays is to facilitate water ingress and softening of the clay, leading to a loss of cohesion through time. After prolonged weathering, an overconsolidated clay may lose all its cohesion by this process. Weathering affects non-fissured clays too, but at a much slower rate. The effects of weathering on clay strength and the formation of hillslopes on clay bedrock is very important and is discussed in 6.2.8. No specific mention has been made of water pressures in clays because implicit in the above discussion is that porewater pressures always exist and are always important in the stability of clay slopes.

4.2.4   Failures in granular soils

The final category of materials, *granular soils*, includes any relatively coarse material (sand size upwards) possessing no cohesion.

Very coarse sediments such as *scree* (Chapter 6.2.7) are impossible to saturate with water and so never experience porewater pressures. Thus their critical stable slope angle is about equal to ϕ, the magnitude of which depends on the surface roughness and shape of the boulders (*see* experiment in 4.2.6). Landslides occur when the slope is steepened above the critical angle, either by erosion at the base or addition of more particles at the top. Typically the slide is a shallow cascade of particles streaming down the slope, which comes to rest at a lower slope angle. *Figure 4.1* shows two slopes where this type of slide is common, a scree slope and a coastal slope composed of coarse glacial sands and gravels.

Cohesionless regoliths accumulating over hard bedrock slopes due to weathering usually contain appreciable fine materials between the larger particles. Thus, porewater pressures can and do occur from time to time, reducing the critical slope angle below ϕ. Characteristically, landslides in this material are shallow, removing the vegetation mat and some of the soil beneath, and leave a bare scar on the slope (*see* example in *Figure 4.1*).

4.2.5 Trigger mechanisms in landslides

We have now examined the force-balance which initiates a landslide and have looked at how material type may influence failure style. There remains one outstanding issue to discuss, namely the question of immediate and long term *causes of landslides*. These are sometimes called *trigger mechanisms* and can be thought of in terms of a long-term background process, overprinted with a number of short-term controls.

The long-term background process is *weathering*. Any material exposed to weathering on a slope loses strength through time, which may ultimately lead to instability. Clearly this is long term and, while helping towards instability, weathering will never be the immediate identifiable cause of a landslide. The major short-term control is *porewater pressure*, which increases after heavy rainfall and is generally higher in the winter when the water table is higher. Thus most slides take place when the ground is at its wettest and

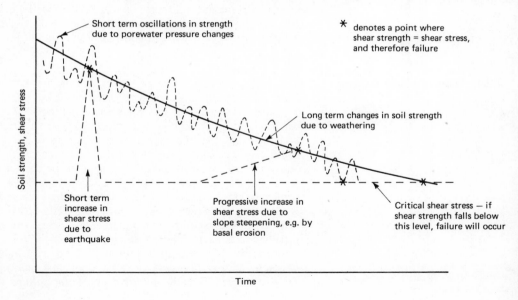

Short term oscillations in strength due to porewater pressure changes

∗ denotes a point where shear strength = shear stress, and therefore failure

Long term changes in soil strength due to weathering

Short term increase in shear stress due to earthquake

Progressive increase in shear stress due to slope steepening, e.g. by basal erosion

Critical shear stress — if shear strength falls below this level, failure will occur

Soil strength, shear stress

Time

Figure 4.8 Long- and short-term trigger mechanisms in landslides

there is a strong association between failure and wet weather. This may be viewed as an annual oscillation of strength superimposed on the general trend due to weathering (*Figure 4.8*). Other short-term triggers include basal erosion of the slope, which causes an increase in overall slope angle until an instability results. This may be due to marine attack on a coastal cliff or lateral undercutting of a valley side by a river. Finally, seismicity in the form of major earthquakes can cause short term strength reduction and hence land-sliding. These last two triggers relate to well defined events with rapidly changing conditions and can cause landsliding on an otherwise long-term stable slope (*Figure 4.8*).

4.2.6 Experiment to measure the critical stable slope angle for a gravel or sand

It is usually very difficult to measure the strength of a soil, and very complex and expensive equipment is necessary. One exception is to measure the strength of dry, cohesionless soils such as gravel or sand, where it is possible to make approximate measurements of the angle of internal shearing resistance, ϕ, using simple makeshift apparatus. The apparatus required consists of a box mounted on a board which can be tilted. *Figure 4.9* below illustrates the important details, although personal modification can be made. A device to measure slope angle is also needed. This could be a protractor mounted on the base of the apparatus (*see* diagram) or a clinometer.

Half fill the box with the sand or gravel to be tested and level the surface until it is parallel with the edge of the box. Tilt the board *slowly and gently* until sliding begins and then measure the board's inclination (α_1). It is helpful to include some kind of locking device to retain the board firmly in position while the angle measurements are made. Now measure the inclination of the sand surface (α_2) after it has come to rest. Record both angles.

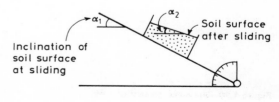

Figure 4.9 Experimental setup to measure critical stable angles for dry gravel and sand

Box with soil

Tilting board

Protractor

Base

Inclination of soil surface at sliding

α_1

α_2

Soil surface after sliding

This illustrates the very important point that there are *two* critical angles, one at which sliding begins and one at which it ceases (the upper and lower angles of repose). The upper angle is approximately equal to ϕ, the angle of internal shearing resistance for the undisturbed sand (not exactly, because ϕ is measured in very precisely controlled conditions). The lower angle is approximately the angle of internal shearing resistance for the disturbed sand, sometimes called the residual angle (ϕ_{res}). The test should be repeated several times to obtain reliable mean angles.

You could repeat the experiment for different materials, for example sand, angular gravel and rounded gravel to show how sediment properties influence critical stable slope angle. In this case you should carefully record grain size and grain shape, and some simple techniques for this may be found in another book in this series (Briggs, 1977). Alternatively you could look at the relationship between measured critical stable angles and slope angles for simple materials such as scree slopes or sand dunes. Field measurement of slope angle is described in Chapter 6.1. Do the maximum slope angles measured on the screes of active, unvegetated sand dunes agree with either of the critical angles measured in the laboratory? If so, do you think that landsliding occurs on the steep slopes? Note that this project is only feasible for effectively dry slopes, of which screes and active sand dunes are the best examples.

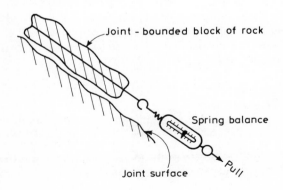

Figure 4.10 The measurement of rock friction

4.2.7 Experiment to measure rock surface friction

It is very simple to measure the frictional strength of rock surfaces and it may be done in the field or laboratory. The only apparatus required is a spring balance and a clinometer. For field measurement, choose a bedding plane or joint surface with a reasonably constant angle. It must not be so steep that blocks of rock will not stay put. Then find a loose block of rock with at least one more-or-less flat side, and attach it to the spring balance with a loop of cord. Place the loose block on the chosen rock surface and pull it down the slope with the spring balance (*Figure 4.10*). Pressure must be applied very slowly and the maximum spring balance reading taken just before sliding starts. Repeat until a reliable mean spring balance reading can be obtained. Then weigh the block and measure the inclination of the rock surface. The angle of plane sliding friction, ϕ_μ, for the block on the surface is then easily calculated as below:

At sliding, the force resisting motion downslope is

$$W \cos \alpha \tan \phi_\mu$$

and the forces promoting movement are

$$F + W \sin \alpha$$

These are equal at sliding, and so

$$W \cos \alpha \tan \phi_\mu = F + W \sin \alpha$$

and therefore

$$\tan \phi_\mu = \frac{(F + W \sin \alpha)}{W \cos \alpha}$$

where W is the weight of the block being towed, α is the inclination of the rock surface, and F is the mean spring balance reading.

The procedure in the laboratory is exactly the same, except that two loose blocks with flat (joint or bedding plane) sides are needed. One block is then towed over the other in a similar manner to above. If the experiment is carried out on a horizontal bench surface, the calculation is even simpler:

$$\tan \phi_\mu = W/F$$

The experiment could be repeated for different rocktypes to show how the frictional strength of joint surfaces may vary with rocktype. At all times, careful observations on the character of the joint or bedding plane surfaces must be made (for example their irregularity, surface mineral coatings, and so on).

4.3 MOVEMENT OF SLIPPED DEBRIS AFTER FAILURE

Usually, debris involved in rapid mass movement only needs to move a short distance downslope to regain stability; or in the case of a fall from a steep slope, to the base of the slope. But in certain circumstances the failed debris becomes mobilised and continues to move for some distance. Usually movement is aided by very high water contents and high porewater pressures, rendering the material very weak and fluid-like in appearance. Accordingly, these movements of wet debris subsequent to failure are known as *flowslides.* The term implies that flow takes place in the debris, though this is not always the case. Many flowslides, despite their fluid appearance, are slides in the strict sense since practically all movement takes place on a single plane at the base of the debris. All types of debris from clay to rock debris may form flowslides, though finer-grained material tends to be more readily mobilised.

4.3.1 Mudslides

After a rotational slip has occurred in clay, and particularly on coastal cliffs where the failed debris and cliff are devoid of vegetation, weathering processes may begin to break down material exposed on the *back scar* (the exposed part of the sliding surface at the top of the slide (*Figure 4.1*)) of the slide. The weathered, loose debris absorbs rain water very readily and slips from the backscar to accumulate as a high water content, muddy deposit on the lower slope (*Figure 4.1*). When sufficient material has accumulated, it begins to move downslope, even on quite low angled slopes of 3–4°, and may develop into a well defined lobe of mobile debris known as a *mudslide*. A typical mudslide is illustrated in *Figure 4.1*. Mudslides often exist for long periods of time with a steady supply of weathered material at the top, balanced by a comparable rate of erosion by the sea at the foot. Superficially, mudslides resemble flows because of their high water content (they are often so wet that it is quite impossible to walk across them) but measurements on them of movement with depth have shown clearly that practically all movement is by sliding on a basal shearplane. Rates of movement are low, a matter of a few metres a year, and practically all of it occurs in the winter months when the debris is at its wettest. When the slide is active, much of the movement takes place as surges, separated by periods of standing still. For mudslides to remain permanent features on a slope for a long period of time, the rate of basal erosion by the sea must only be moderate, so that supply of debris to the slide remains in balance with removal. In the London Clay cliffs of the Thames Estuary, mudslides are a common process where rates of cliff retreat are at about 0.3–0.8 m yr^{-1}, but are replaced by deep seated rotational slips at higher rates. At high rates of retreat there is not enough time for sufficient accumulation of weathered debris between rotational slips for permanent mudslides to develop.

4.3.2 Debris flows

Coarser materials are much less easy to mobilise than fine, clay debris although flowslides in this type of debris are by no means uncommon. One fairly common type of flowslide in coarse debris is the *debris flow*. A debris flow consists of a lobe of very wet debris, typically about a metre across and a few cubic metres in volume, which moves down a hillslope at a few kilometres an hour. As it progresses, it tends to leave banks of material, or *levées*, on either side of its path (*Figure 4.1*). The levées seem to develop because boulders in the flowing debris tend to block the forward progress of the flow, and as pressure builds up behind they are 'snowploughed' aside to form the banks. Debris flows often occur in scree-like material, especially when some fine material is present between the coarser particles to retain water. They typically start to move on slopes of 30–40° and continue to flow over slopes of 12–20°.

On lower slopes they decelerate and eventually stop. Whether they are flows or slides is unclear since very few observations have been made on moving debris flows. However, they are undoubtedly associated with very high water contents. They commonly occur in scree slopes where springs rise beneath them, and during very heavy rain along gullies cut through scree slopes. They operate in a wide range of climates from semi-arid to cold-alpine; indeed anywhere high rainfall intensities occur. Their influence on slope-form is taken up in Chapter 6.3.

Only two fairly commonly occurring types of flowslide have been described in this section although many other styles do exist. They often occur in spoil tip material, which resembles scree in many ways, and frequently have serious consequences. This aspect of flowslides is taken up in 7.1.6. The precise nature of a flowslide depends upon the mechanical properties of the material involved and the topographic situation, and it is difficult to generalise about this extraordinarily diverse group of processes.

4.4 SLOW MASS MOVEMENT

The rapid mass movement processes that have just been described all have at least one common feature: the moving mass is separated from the stationary hillside by a shear plane. With the slow processes there is no such clear division between moving and stationary material; the movement simply dies away with depth in the soil or regolith. *Creep* in soils is commonly defined as *the slow downslope movement of superficial soil or rock debris, usually imperceptible except to observations of long duration*. This definition was first proposed in 1938 and since that time instruments have been developed which can detect very small movements, so that it is now possible to measure creep over time periods of a month, or even less in some cases. It is perhaps more appropriate therefore to think of creep in soils as a slow downslope movement of soil or regolith which is not bounded by a shear plane. Shearing must occur during creep but it is distributed as small movements throughout the soil mass.

In the classification shown in *Figure 4.1* the slow mass movement processes are divided into two groups, *soil creep* and *solifluction*. The basis of this division is the relative importance of freezing and thawing. Freeze–thaw is a contributory factor to soil creep, given favourable climatic conditions, but it is the major factor in solifluction. A continuum exists between areas where freeze–thaw is absent (e.g. the humid tropical lowlands) to areas where it is predominant (e.g. subarctic climates) so these processes are also connected by a continuum.

Soil creep is considered to include three major groups of processes, *seasonal creep*, *continuous creep* and *random creep* and the distinction between them is made on the basis of the forces generating movement. Gravity is the driving force for continuous creep but in seasonal and random creep, movement is produced by a variety of forces which will be discussed below. In these cases gravity is only effective when the material has been set in motion by other forces. It is a relatively simple matter to make theoretical distinctions between types of creep but in practice it is almost impossible to distinguish between them by measurement. Whatever the actual causes of soil creep, the fact remains that it is an almost ubiquitous process and as such is very important in the development of hillslopes.

4.4.1 Seasonal creep

Seasonal creep is caused by seasonal or cyclic processes in the soil, the most important of which is probably wetting and drying, but freeze–thaw and heating and cooling also play a part. With increasing latitude (or altitude) freeze–thaw will become more important until the process grades into solifluction.

The effects of all these cyclic factors is to cause the soil to expand and contract. For wetting and drying cycles the amount of soil movement, for a given moisture content change, will depend on the amount of clay in the soil and the type of clay minerals present. The swelling capacities of the three major types of clay minerals are shown in *Table 3.1* (p. 71). For most soils, moisture changes will be mainly confined to the upper soil horizons where plant root density is highest, since most of the drying of the soil will be due to evapotranspiration by plants. Freezing cycles will be most frequent at the soil surface where diurnal and seasonal temperature changes will also be concentrated. Soil movement by all causes is therefore greatest at the soil surface and diminishes rapidly with depth.

The commonly accepted theory of seasonal soil creep is based on the assumption that expansion of the soil occurs in a direction normal to the soil surface. In expanding, the soil loses some of its cohesive strength so the movement is deflected slightly downslope under the influence of gravity. In contracting, the soil tends to move back normal to the soil surface but is again deflected by gravity so that any individual soil particle will follow a ⌒-shaped path during an expansion and contraction cycle. This is probably not the case for movements generated by freeze–thaw but these will be discussed under solifluction below. The theoretical velocity profile of movement produced by moisture changes is shown in *Figure 4.11*.

By this theory, all the observed movement caused by seasonal soil creep processes should be downslope and the rate of movement should be proportional to the sine of the slope angle. In practice, neither of these has been found to be the case for the short periods of record which, at present, are the only ones available. Since soil properties are important in determining how much movement will occur and these vary quite widely

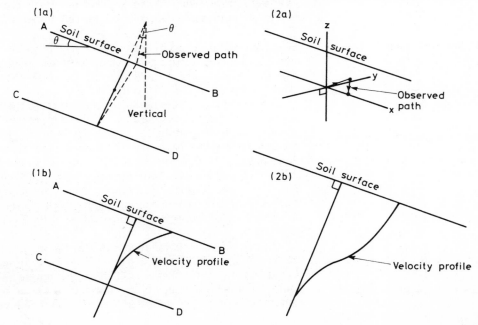

Figure 4.11 (a) Theoretical particle movements and (b) velocity profiles for (1) freeze–thaw and (2) moisture content change in soils

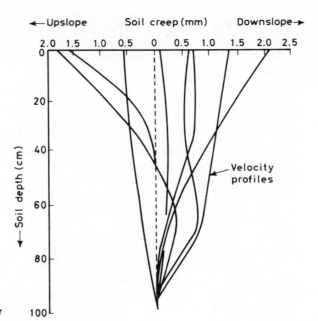

Figure 4.12 Velocity profiles of soil creep measured in the Mendip Hills (Somerset, UK) over a period of one year

over even small areas of slopes, it is likely that they tend to obscure any relationship between creep and slope angle. The basic theory, that creep rate will be proportional to the sine of the slope angle, *all other things being equal*, is acceptable but in practice there is sufficient random variation to obscure the relationship. The other, more difficult, problem is that velocity profiles of creep, especially when measured over one or two years or less, often go upslope. The velocity profiles shown in *Figure 4.12* were all measured over a period of one year in a small area on the Mendip Hills (Somerset, UK) and a number of upslope movements occurred. Many similar results have been reported and they suggest that the theory is wrong in assuming that expansion and contraction only operate normal to the soil surface. In fact, lateral movements must also occur as can be seen from the cracking of soils on drying. Seasonal soil creep is not yet fully understood, and there is a need for more field measurements, especially of long term movements, and more theoretical work.

4.4.2 Continuous creep

Unlike seasonal creep, the movement of soils by continuous creep is entirely due to the force of gravity. However, just as for seasonal creep, continuous creep depends on the presence of a significant clay fraction in the soil. Below a critical level of applied stress, sometimes called *fundamental strength*, no deformation (strain) will occur in a clay. Clays also possess another critical level of strength called the *yield strength*, and only if this is exceeded will failure occur. At stress levels between the fundamental strength and the yield strength of clay, creep will take place. The stress is produced by overburden pressure so that continuous creep will only occur at some depth below the soil surface. The cohesive strength of clay decreases with increasing moisture content and continuous creep will therefore be fastest when moisture content is high. Since moisture content varies seasonally it follows that the rate of continuous creep will also vary seasonally.

On some slopes, continuous creep occurs in a narrow band of the soil and this eventually leads to a reduction in shear strength in that band by reorientation of the clay particles. Thus it is not uncommon to find that some rapid mass failures have been preceded by a period of continuous creep in the zone which becomes the shear plane during failure.

4.4.3 Random creep

Whenever soil particles are set in motion on a slope the force of gravity will cause them to come to rest in a position slightly downslope of their starting point. This is the phenomenon which we call *random creep*. The construction and collapse of animal burrows, the growth and decay of plant roots, the movement of soil by the roots of falling trees (tree-throw), trampling by humans and animals and a great variety of similar processes all contribute to random creep. Rainsplash can also be included with random creep although here we have chosen to include it in Chapter 5 as a fluvial process.

4.4.4 Solifluction

The term *solifluction* is used to cover a range of processes in areas where the soil freezes during winter and thaws in the spring, and where permafrost (permanently frozen ground) may be present. At one end of the spectrum solifluction is very similar to seasonal creep (*see Figure 4.1*) where movement is generated by alternate freezing and thawing of soil water and is limited to the depth of penetration of freezing. At the other extreme, solifluction may resemble a mudslide where thawing ice-lenses increase porewater pressures, especially over permafrost, and the soil slides downslope over a shear plane. Rates of movement by solifluction are commonly much greater than seasonal creep rates but are concentrated into a much shorter time period. Movement is also often locally concentrated in solifluction lobes between areas of more stable ground.

4.4.5 Measurement and observation of soil creep

Although it is possible to distinguish in theory between different creep processes, it is difficult, or even impossible, to do so in practice. It appears likely that seasonal creep is quantitatively the most important process because it is so widespread but it is not really possible to prove this experimentally. Perhaps the biggest problem with soil creep is the difficulty of making reliable field measurements. Reference points are established in the soil and their positions determined instrumentally at various times, to give velocity profiles of movement with respect to some point, at depth in the soil (or in the bedrock), which is assumed to remain stationary. Most of the techniques devised so far involve a considerable amount of disturbance of the measuring site and it is difficult to establish just how much of the observed movement is due to disturbance and how much to soil creep. A simple technique is described in 4.4.6, but great care is needed in carrying it out because of the disturbance involved. The method of calculating movement and the units of measurement used are also described in 4.4.6.

A number of features on slopes are often taken to be indicators of the action of soil creep. For example, it has often been observed that tree trunks are curved upslope, indicating that creep is tilting the trunk downslope and that vertical stem growth compensates against this, producing curvature in the stem. While this is an appealing idea, the observed facts do not entirely support it. Tree trunks in any area often curve in a variety of directions and curvature is usually concentrated at the base of the trunk. If the process described above were the cause of curvature, the whole trunk should follow a smooth curve, convex downslope. Similarly posts and fences are often observed to be leaning downslope and this is also assumed to be due to creep. While creep may be involved, this phenomena can also be explained in terms of the increased loading on the soil imposed by the posts themselves. More convincing indicators of soil creep are displaced fragments of bedrock in a soil. This is often particularly obvious where the bedrock is slate or shale and the lineations in the bedrock can be seen in section to curve away downslope in the soil profile.

4.4.6 Experiments to measure soil creep

The technique to measure creep movements described here is known as a Young Pit. A pit is dug in the soil across the slope so that one narrow face is aligned at right angles to the contours. If possible the pit should be dug down to bedrock. A stout steel stake is driven into the bedrock, its top inscribed with a cross (+) and covered with rust-proofing paint. A series of markers are then inserted into the face of the pit above the stake (*Figure 4.13*).

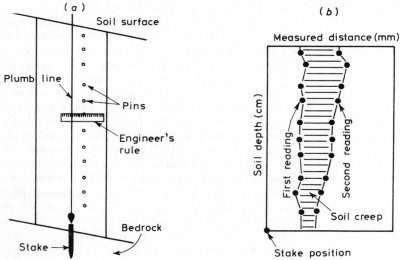

Figure 4.13a and b Method of taking measurement in a Young Pit and calculation of volumetric creep from two sets of readings

The markers should be of fairly small diameter (2–3 mm) and about 5 cm long, brightly coloured, and made of a non-ferrous metal such as brass or aluminium. A plumb-line is then suspended over the stake with its point at the cross on the top of the stake. The distance between the plumb line and each marker in the pit face is then carefully measured with a good quality engineer's rule. Care must be taken to avoid parallax error. The height of each marker above the stake should also be recorded. As many markers as possible should be used. If possible, have a number of people take the measurements independently of each other and obtain mean distances from their results.

Before filling in the pit place a sheet of newspaper over the face containing the markers. Carefully fill the pit, being sure not to disturb the markers, and try to pack the soil to its original density. Accurately record the location of the pit with reference to a number of fixed points (e.g. fence posts). Measure the slope angle at the pit.

After a suitable time has elapsed, (at least a year but preferably more), reopen the pit by digging carefully towards the marked face. The newspaper will allow the face to be uncovered with the minimum of disturbance. Remeasure all the pins, preferably with the same rule, and refill the pit so that it can be used again.

There are now two measurements for each marker and the difference between them is the distance the marker has moved relative to the stake. Plot these differences on graph paper against the height of the marker above the stake as shown in *Figure 4.13*. The shaded area represents the amount of soil creep that has occurred. Do not be surprised if some or even all of the markers have moved upslope! The area can be measured by counting squares on the graph paper and this should be expressed in square centimetres. Assume that the observed movement applies to a one centimetre width of slope and you now have the soil creep rate in cubic centimetres per centimetre (cm^3 cm^{-1}).

Divide the cm^3 cm^{-1} by the number of days between the readings and multiply by 365 to give the annual creep rate in cm^3 cm^{-1} yr^{-1}.

Note that much more accurate measurements of the marker positions can be made with a theodolite, if one is available.

CHAPTER 5   WATER ON HILLSLOPES

5.1 THE HYDROLOGY OF HILLSLOPES

In order to be able to understand the erosion of hillslopes by water it is necessary to begin by considering their hydrology. The behaviour of water on hillslopes is governed by two fundamental hydrological principles. The first is based on the simple fact that water cannot be created or destroyed. It changes state, between solid (ice), liquid, and gas (water vapour), and it changes location. It is these changes that are the subject matter of the hydrological cycle. This principle is best expressed by the *universal hydrological equation*:

$$\text{Input} = \text{Output} \pm \text{change in storage}$$

For any given space, for example a hillslope, this equation always balances. The second basic principle is that water always flows down an energy gradient. In this context this usually means downhill, especially under saturated conditions (*Figure 3.1*). Under certain unsaturated conditions (*Figure 3.1*) it may move physically uphill due to capillary tension, though this will still be down an energy gradient.

5.1.1   The hydrological cycle

The major components of the hydrological cycle on a soil covered hillslope are shown in *Figure 5.1*. This shows a 'slice' of a hillslope from the divide to the stream across a section of parallel contours, such that the only inputs of water are from precipitation. Depending on the particular climate, precipitation will consist of varying proportions of rain, hail, sleet, snow and direct condensation. In the following discussion only precipitation as rainfall will be considered but, with slight modifications, it would apply equally to precipitation in any form.

In most areas, and especially the humid climates, much of the rainfall will land on the vegetation cover and be stored on leaf and stem surfaces. Notice how the pavement underneath trees remains dry for some time after the start of rainfall and how much more pronounced this is in summer, when the trees are in full leaf, than in winter. Rainfall caught in the vegetation is called *interception storage* and when this storage capacity is

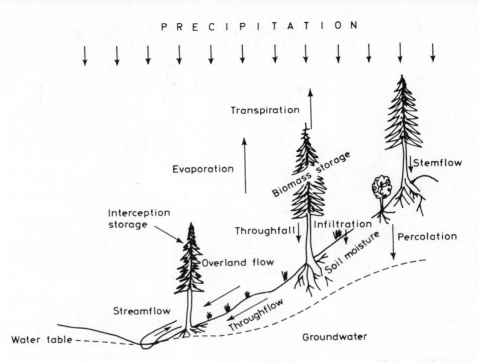

PRECIPITATION

Transpiration

Evaporation

Biomass storage

Stemflow

Interception storage

Throughfall

Infiltration

Percolation

Overland flow

Soil moisture

Streamflow

Throughflow

Water table

Groundwater

Figure 5.1 The hydrological cycle on a hillslope

full some will fall on to the ground (*throughfall*) and some will run off down plant stems (*stemflow*). Some of the rainfall is evaporated back into the atmosphere directly from interception storage.

Throughfall and stemflow, together with rainfall that reaches the soil surface directly, will then take one of three paths. It may remain in depressions on the soil surface as *surface storage*; it may flow downslope on the soil surface as *overland flow* (also called *surface runoff*); or it may *infiltrate* into the soil. The rate at which water can infiltrate into a soil is its *infiltration capacity*. This is a very important soil property which is discussed in more detail below and a simple technique for measuring it is described in 5.1.4. Water which infiltrates into the soil initially fills the pore-spaces of the soil and then either *percolates* vertically downward into the soil and ultimately to the groundwater table, or flows laterally downslope through the soil as *throughflow* (also called *interflow*).

Water is lost from a hillslope in a number of ways, some of which have already been mentioned. Loss of water can occur by percolation into the groundwater, by overland flow and throughflow into the stream at the base of the slope, or by the processes of evaporation and transpiration which are collectively known as *evapotranspiration*. Water is stored on slopes in surface depressions (*depression storage*), in soils where it is referred to as *soil moisture*, and in the tissues of plants (*biomass storage*). The important point, already made above, is that the total amount of water being distributed in these various ways equals the total precipitation. The universal hydrologic equation when applied to a hillslope becomes (*see Figure 5.1*):

> Precipitation = Overland flow + throughflow + percolation to groundwater + evapotranspiration ± change in depression, soil and biomass storage.

For erosion by water to occur on a hillslope, water must flow from the slope. Climate therefore has an important effect on hillslope erosion by controlling not only the amount of precipitation received, its intensity, and its seasonal distribution, but also the amount of evapotranspiration which will occur. The larger the proportion of precipitation that is lost in evapotranspiration, the less there will be available for runoff and erosion (Chapter 5.2.2).

The total amount of erosion on a slope therefore depends on the total amount of runoff while the type of erosion (i.e. mechanical or chemical) depends on the type of runoff. The processes involved will be discussed in Chapters 5.2 and 5.3. The production of overland flow is governed by the infiltration capacity of the soil and the intensity and duration of rainfall. Overland flow, and therefore mechanical erosion, are governed by conditions prevailing at the time the rainfall occurs and by changes in these conditions throughout the period of rainfall. Throughflow and percolation, and therefore chemical erosion, are more closely related to the longer term balance between precipitation and evapotranspiration and the seasonal distribution of precipitation.

5.1.2 Infiltration capacity and flow through soils

Infiltration capacity

The infiltration capacity of a soil is a measure of the rate at which water can pass into the soil through its surface, and is usually expressed as millimetres per hour, the same units as are used to measure rainfall intensity. All soils have a characteristic infiltration curve through time (*Figure 5.2*), with infiltration rate high at the start of wetting and falling off progressively as wetting continues. Fine-grained (especially clay) soils tend to have lower infiltration capacities than coarser grained soils and on any given soil the infiltration capacity will vary with vegetation type. Thus infiltration rates tend to be highest under a forest cover, lower under grass and lowest of all where the soil surface is bare. The soil moisture content at the start of wetting also influences infiltration capacity, with dry soils having higher rates than wet ones.

Overland flow

Since infiltration capacity falls with time from start of wetting, if the rainfall persists long enough the infiltration capacity may fall below the rainfall intensity and **infiltration excess overland flow** will occur. A rapid decline in permeability occurs at the soil bedrock interface and in some soils there is a layer or **horizon** at shallow depth in the soil which inhibits water movement. Podsol soils, which have a middle horizon enriched with deposited iron minerals and clay particles leached from above, are a good example of soils where permeability decreases with depth. The soil may fill with water above such layers, leading to **saturation overland flow.**

In humid climatic areas, for example the British Isles, the infiltration capacities of most soils are quite high so that infiltration excess overland flow rarely occurs. It is a far more common occurrence in arid and semi-arid areas. Where overland flow does occur in humid areas it is usually as saturation overland flow, commonly on sites already quite moist at the start of rainfall. While the amount of overland flow is controlled by the infiltration process, its speed and therefore erosive power is primarily controlled by slope. Water infiltrating

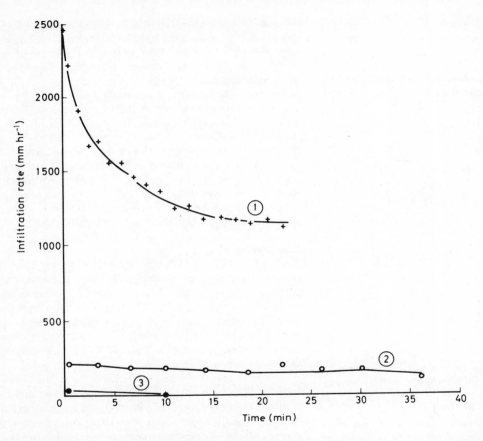

Figure 5.2 Measurements of infiltration rates under forest and pasture made in the area shown in *Figure 5.3*

into the soil will be stored as soil moisture or will flow downslope as throughflow or percolate to the groundwater. The velocity of throughflow is governed by the slope and *permeability* of the soil. Permeability is a measure of the rate at which water will flow through the soil and is controlled by the size, interconnectedness, and tortuosity of the pore-spaces. Generally speaking, permeability is low in fine grained soils and increases as the grain size of the soil increases.

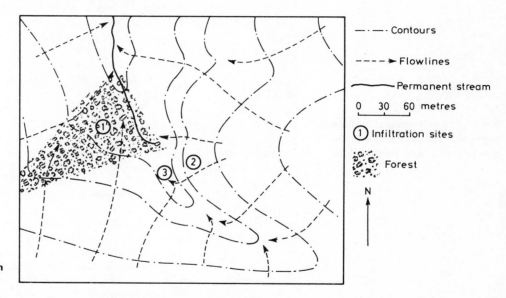

Figure 5.3 Part of a small drainage basin near Grendon Underwood (Bucks, UK)

Water both in and on the soil flows downslope along lines which cross all the contours at right angles. A typical pattern of *flowlines* is shown in *Figure 5.3*. This pattern of movement leads to consistently higher soil moisture contents at the base of the slope and where the flowlines diverge on ridges or spurs.

Given the relationship mentioned above between soil moisture content and infiltration capacity, overland flow will tend to occur mainly in those areas having consistently higher soil moisture contents. This pattern is illustrated by a series of infiltration measurements made at the sites shown in *Figure 5.3* near Grendon Underwood in Buckinghamshire, UK. Infiltration rates were measured as described in 5.1.4 and the results are shown in *Figure 5.2*. At site 1, under a forest cover, the infiltration rate shows the characteristic decrease through time but never falls to below the rainfall intensity which could be expected in this area. Site 2 has a grass cover but is on a water shedding site. The infiltration rate is lower than under the forest but still well above rainfall intensity. At site 3, a hollow and therefore an area of converging flow lines and higher soil moisture content, the infiltration rate is very low and could easily be exceeded by rainfall intensity. Overland flow in this area has been observed to occur only in the hollows, as would be expected from the pattern of infiltration rates.

5.1.3 The influence of climate and vegetation on the balance between overland flow and throughflow

In humid regions, particularly humid temperate areas such as the British Isles, there is sufficient available moisture to support a complete vegetation cover. The vegetation protects the soil beneath from the direct impact of raindrops and also reduces the total amount of rainfall reaching the ground (owing to interception storage). In addition, the vegetation litter and network of roots help to maintain a high infiltration capacity in the soil, especially under forest (*see Figure 5.2*). The end result of the combination of all these factors is that in humid areas, overland flow is relatively rare and occurs mainly at the slope base or in hollows where the soil is usually wet at the beginning of rainfall (i.e. situations such as described in 5.1.2). Under such conditions, throughflow in the soil is the most important mode of downslope water movement.

By contrast, in arid and semi-arid areas there is insufficient moisture available to maintain a complete vegetation cover and bare soil is exposed directly to raindrops. Rainfall on to bare soil tends to splash fine grained material into soil pores and cracks, thus sealing them and lowering the infiltration capacity. Many desert soils have surface salt crusts deposited when soil water evaporates and these also help to seal the soil surface. Given that rainfall intensities tend to be relatively high in these areas, the infiltration capacity of the soil is often exceeded all over the slope and widespread overland flow occurs. Water which does infiltrate will be evaporated from the soil or transpired by plants so that little, if any, travels downslope as throughflow.

In semi-arid and arid areas therefore, runoff is predominantly as overland flow and throughflow is very limited. This is in contrast to humid climatic areas where overland flow occurs only in restricted locations and is a very small proportion of the total runoff (usually less than 10%).

5.1.4 Field measurement of infiltration capacity

Equipment: A number of 15 cm lengths of 10 cm diameter plastic drainpipe, bevelled from the outside at one end (infiltration rings), block of wood, hammer, large jerrycan of water and a ruler.

Procedure
Drive the sharpened end of the infiltration ring squarely into the soil by placing the block of wood across the top and hitting it with the hammer (*Figure 5.4*). It is important that

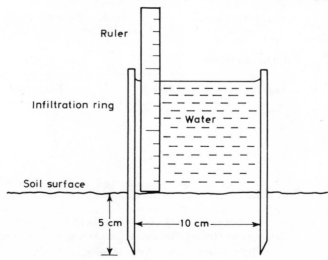

Figure 5.4 Experimental setup for measuring infiltration rate.

117

the ring is driven in squarely so that the soil inside it makes a firm fit with the walls of the pipe. Drive the ring to a depth of 5 cm. Two records of time must be kept during the experiment and will be referred to here as time A and time B. Time A is a record of time elapsed since the beginning of the experiment. Start the experiment by filling the ring with water and record time A as on the form shown below. Place the ruler inside the ring so that you can observe when the water level has fallen 1 cm. The time taken for a fall of 1 cm is recorded as time B. Refill the ring quickly and again record time A. repeat the procedure for a fall of 1 cm until the length of time taken for a fall of 1 cm (time B) becomes relatively constant. This may take anything up to two hours or even longer. Calculate the infiltration rate in millimetres per hour for each measured fall as:

$$\frac{\text{Fall in water level (in millimetres)}}{\text{Time B (in hours)}}$$

and plot these values on a graph (as in *Figure 5.2*) with time A on the *X*-axis and infiltration rate on the *Y*-axis. The results can be set out as in the table shown.

RESULTS FORM

Time A	Time B	Fall in level	Infiltration rate

A fall in water level of 1 cm has been recommended here as being a reasonable value to use but may have to be varied depending on site conditions. Where the infiltration rate is very high it may be necessary to use a fall of 10 cm between refills whereas where the infiltration rate is very low it may be necessary to time the fall over only a few millimetres. The procedure will have to be adapted to suit local conditions.

5.2 MECHANICAL EROSION BY WATER

5.2.1 The geomorphic significance of slope hydrology and vegetation

In 5.1.3 the very important links which exist between soils, vegetation and hydrology were discussed. The relationships which exist between all these parts of a hillslope system are of the greatest importance to the geomorphic processes discussed in the remainder of this chapter, and they are stressed throughout.

Vegetation cover and soil characteristics largely control the broad hydrology of a slope, in that they determine where the water goes. Typically, well-vegetated soils have a good, open structure, are therefore quite permeable and tend to favour infiltration. Water therefore moves predominantly within the soil as throughflow. The reverse is true of bare or poorly vegetated soils. Their permeabilities are usually low, and water moves off them mainly as overland flow. A broad, and obviously generalised, relationship exists between soil–vegetation characteristics and slope hydrology. These important interactions have implications for processes operating on slopes.

Overland flow has a high velocity, usually in the range of $0.1–1$ m s^{-1}, and is capable of entraining and transporting sediment particles from the slope. This is usually known as slopewash. At the same time it spends only a short time in contact with the sediment on the slope and has little opportunity to dissolve material. Thus, overland flow is characteristically high in sediment load but low in dissolved load. In contrast, throughflow has a low velocity ($0.001–0.1$ cm s^{-1}), and therefore has a small entrainment capacity. It does remain in contact with the soil for a long time, however, and has ample opportunity to dissolve a considerable chemical load (Chapter 5.3). One may therefore extend the simple generalisations made here to say that low infiltration capacity soils are characterised by sediment loss due to overland flow (slopewash), whereas high infiltration capacity soils have low rates of surface erosion but high rates of subsurface chemical erosion. While this is obviously a rather simple view, it provides a useful rule-of-thumb guide to the dominance of certain processes on slopes.

The importance of the soil–vegetation system in controlling erosion by water on slopes extends beyond influencing the simple ratio between overland flow and throughflow. For example, on well-vegetated slopes overland flow is retarded owing to frictional drag in the vegetation layer, and its capacity for erosion is reduced. Also, a close vegetation mat will effectively prevent intimate contact between the flowing water layer and the mineral soil, reducing surface mechanical erosion to very low levels. Thus, not only does less water flow overland on well-vegetated slopes, the ability of what flowing water remains to erode is also reduced. In a similar fashion, vegetation prevents the full impact of raindrop erosion being realised at the ground surface. We shall also see below that the presence of vegetation is important in enhancing subsurface chemical erosion by providing the source of complex organic acids, which have a major part to play in mobilising certain chemical constituents within the soil (Chapter 5.3.2). Once again, the effect of vegetation in controlling chemical erosion goes beyond simply increasing throughflow; it actually plays a part in the chemical processes.

5.2.2 Climate and water erosion on slopes

From the above summary of the interplay between soils, vegetation, hydrology, and geomorphic processes, it should be apparent that there are possible links between climate and hillslope processes which require some exploration. Bare or poorly-vegetated slopes, the sites of significant slopewash erosion, are really only important in areas where rainfall is too low to support a permanent or continuous vegetation cover. These areas are usually where evaporation potential is close to, or in excess of, rainfall. Hence, semi-arid and arid areas are most susceptible to mechanical erosion due to overland flow.

In the humid—temperate climate enjoyed in the British Isles, naturally occurring bare-soil slopes are rare and usually temporary. Consequently, slopewash is of subsidiary importance. Even on slopes too gentle to be subject to landsliding, it has been estimated that soil creep may be more than ten times faster than slopewash under current climatic conditions operating in Britain. Naturally, this is not necessarily true of slopes which have artificially been laid bare. Motorway cuttings frequently show signs of surface water erosion until vegetation becomes established, and agricultural land is subjected to accelerated erosion when left bare after cropping (Chapters 1.3.2, 7.2 and 7.1.2). The latter is not often a problem in Britain (though exceptions occur) but is extremely problematic over much of the world, especially in drier and more marginal conditions.

Bare ground controls the susceptibility of a slope to slopewash but it does not solely control the importance of the process. Obviously rainfall is a major control and the two characteristics which describe it, **_rainfall intensity and total rainfall_**, require discussion. Rainfall intensity is the rate in mm hr^{-1} at which rain falls, and higher intensity rains have greater erosive capacity. One of the reasons that plants soon colonise newly exposed bare ground in Britain is that rainfall is not usually intense enough throughout the year to prevent rooting of seedlings by erosion. The second factor, total rainfall, is the control on the maximum potential for slopewash, since it is generally true to say that overland flow increases with rainfall. Hence, rates of overland flow and slopewash from bare ground

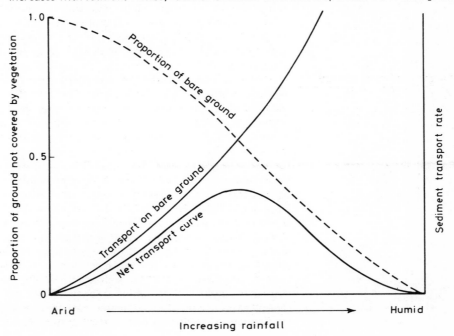

Figure 5.5 The influence of climate on slopewash

areas should increase as rainfall increases. Putting together the relationships of proportion of bare ground to rainfall and rates of slopewash erosion to rainfall gives us *Figure 5.5*. The figure shows that if we combine the two relationships by multiplying the slopewash erosion rate by the proportion of bare ground, a peak of erosion rate occurs in the semi-arid climatic zone. The predominance of slopewash in semi-arid environments is reinforced by the fact that what rainfall does occur tends to be of high intensity, and is therefore more erosive.

To summarise this simple discussion of the complex problem of water erosion on slopes, one may say that regions of bare and low-permeability soils (usually dry climates) are dominated by slopewash, and that well-vegetated and high permeability soils (usually humid climates) are more subject to subsurface chemical erosion. The importance of this with respect to the development of slopes in different climates is discussed in Chapter 6.4.

The remainder of this section is concerned with the processes of surface and subsurface mechanical erosion by water on slopes. Surface water erosion is a complex group of processes in which rainsplash, unconcentrated overland flow, and concentrated flow in small channels or *rills*, all combine to move sediment. They operate together, though it is convenient to view them as distinct processes for the purposes of explanation.

5.2.3 Rainsplash

Raindrops striking the ground surface have a capacity to erode soil which depends on their impact energy. Impact energy increases with size and velocity of the raindrops and both of these tend to increase with rainfall intensity. On well-vegetated ground, already mentioned above, the vegetation absorbs practically all of the rain's energy and so no erosion takes place. But on poorly vegetated ground, where the drops strike the surface with little interruption, considerable erosion may result. In addition to erosion, rainbeat tends to compact bare soil and to reduce its permeability, leading to an increase in overland flow, and therefore slopewash.

Impacting raindrops tend to shatter into tiny droplets which rebound in parabolic flight paths away from the point of impact. They carry small particles of soil with them, causing a redistribution of soil material. Raindrops also push soil particles up to about 50 mm diameter by direct impact and can move much larger fragments indirectly by

Figure 5.6 Earth pillars protected from rainsplash by capping stones

undermining them. *Figure 5.6* illustrates the latter very well. On an area which has been subject to rainsplash erosion, the larger stones stand on small pedestals of soil which have been protected from direct impact. The height of the pedestals gives some idea of how much splash erosion has taken place. The experiment in 5.2.6 describes a technique to allow you to demonstrate the ability of rain drops to perform this type of erosion.

Figure 5.7 summarises the effect of impacting raindrops. Compaction, splashing and pushing all result from raindrop impact. On level ground the net effect of splashing and pushing is equal in all directions and no net erosion takes place. But on a slope, the splash trajectories have a much longer downslope path than upslope and a component of the impact force will act downslope to push soil grains. A net transport of soil results from the impact process, which should increase in importance with slope angle. In laboratory simulations of rainsplash, where artificial rain was allowed to fall on a tray of soil, it has been shown that the net rate of slopewash transport downslope increased by six times as the slope angle was raised from 5° to 25°. Further, the experiments showed that on zero

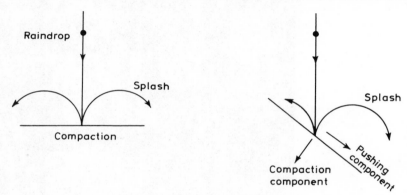

Figure 5.7 Forces involved in rainsplash transport

slope the amount of splash was equal in all directions, but as slope angle was increased to 25° over 95% of the total splash was in the downslope direction. You can demonstrate the importance of slope angle on rainsplash yourselves by performing the experiment in 5.2.6. Field studies of rainsplash have not demonstrated the influence of slope angle as convincingly as the laboratory models, because so many other factors can vary in the field situation.

As rainfall continues, the direct impact of falling drops is reduced as overland flow is generated. A film of flowing water tends to cushion the soil against the impact force of falling rain and so rainsplash is only directly responsible for significant sediment movement where overland flow is of little importance. It has an indirect effect, however, in that it increases turbulence in the flow, as discussed below. The important area where overland flow is small is close to the top of the slope, where little catchment exists to generate flow. Thus rainsplash becomes an important slope-forming process near the slope crest, a point which is discussed further in Chapter 6.3.

5.2.4 Erosion by overland flow or 'sheet' erosion

Water flowing over a slope has the capacity to erode soil when the shear stress applied to the soil layer exceeds the forces of resistance tending to hold the grains in place. The process can therefore be comprehended in simple mechanical terms. Resistance to erosion is proportional to the weight of individual grains acting vertically, and to cohesion between grains. The forces tending to cause movement, on the other hand, are several. First, the flowing water exerts a shear stress at its base, tending to roll or push the grains along the

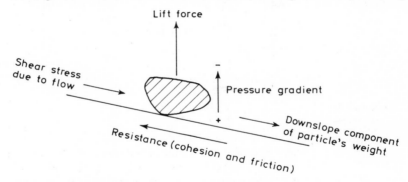

Figure 5.8 Forces acting upon a static particle in flowing water

bed. Second, a fluid flowing past a particle on its bed tends to set up a pressure gradient from base to top of the grain, and causes a vertical lift force to act. Third, the downslope component of the grain's own weight acts to move the particle along the bed. These forces are summarised in *Figure 5.8*.

Flowing water on slopes may be laminar or turbulent. Laminar flow exists when the flowlines in water are parallel to each other and there is no vertical or lateral mixing throughout the fluid. Velocity increases steadily away from the bed (*Figure 5.9*). In turbulent flow vertical mixing does occur, resulting in faster water from the top of the flow being mixed with the slower water near the bed, and vice versa. In the case of turbulent flow, velocity increases much faster close to the bed and the flow is more erosive than laminar flow (*Figure 5.9*).

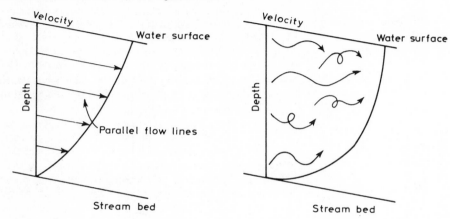

Figure 5.9 Laminar and tubulent flow

Unconcentrated sheetflow on slopes can be laminar or turbulent, depending on the depth and slope of the flow, and upon the amount of turbulence created by impacting raindrops. But if flow becomes concentrated into well-defined flow paths on the slope, it is usually turbulent. Concentrated flow is therefore much more erosive than the slower unconcentrated flow. Calculating the forces involved and predicting when transport will begin in overland flow is a very difficult problem in fluid mechanics and will not be discussed here. The important point to note is that a critical depth of overland flow is required before any transport will take place. Close to the slope crest there is little catchment area and consequently a small depth of overland flow during rainfall. There is therefore little or no capacity for overland flow erosion close to the divide. This is precisely the zone where rainsplash is dominant. The depth of flow increases downslope as the catchment increases and, where the critical depth is attained, erosion begins. Theoretically, the rate of erosion should steadily increase downslope as flow depth progressively increases, though for many reasons this is not always the case. The relationship between overland flow and slope form is discussed further in Chapter 6.3. The rate of overland flow erosion should also increase with slope angle, though once again this is not always demonstrable in the field, because the relationship is obscured by many other factors.

After flow has started on a slope there is always some tendency for it to become concentrated into more-or-less well-defined flow paths. These may be due to minor irregularities on the slope, causing the flowing water to converge at some points but not at others. Water flowing along concentrated routes tends to be much more erosive than unconcentrated flow because of its faster velocity. If the flowpaths deepen sufficiently to become recognisable channels they are called *rills*. Rills may show cycles of formation and destruction from storm to storm, or from year to year. That is, they are formed during a storm or series of storms and tend to be destroyed during the intervening dry periods by weathering processes. New rills may be formed again but in different places, and so the whole slope is affected by rill erosion over a long period of time. If rills persist and deepen progressively through time they become part of the permanent drainage system, with their own valley slopes facing towards them.

Figure 5.10 is a generalised plan of a section of slope undergoing surface water flow erosion. The accompanying photograph shows an example of a typical natural slope where water flow erosion is important, and it is significant to note how poor the vegetation cover is. Unconcentrated flow on slopes generally has only a short distance to travel before it reaches a channel, and so discharge and transport of sediment by unconcentrated flow does not generally increase downslope. Channels are responsible for the vast majority of sediment transport from the slope, and because they slowly change position with time they affect the whole slope. The channels are shown as braided in the diagram, which would be typical in fairly loose unconsolidated slope sediments. If the material is cohesive and less erodible, the channels tend to form better integrated systems, like those in the photograph. The zone of dominant rainsplash is also visible in the diagram and photograph in *Figure 5.10*, as a band of unchannelled hillslopes at the top of the slope.

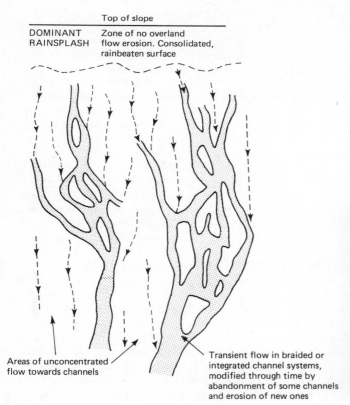

Top of slope

DOMINANT RAINSPLASH

Zone of no overland flow erosion. Consolidated, rainbeaten surface

Areas of unconcentrated flow towards channels

Transient flow in braided or integrated channel systems, modified through time by abandonment of some channels and erosion of new ones

Figure 5.10 A generalised view of slopes undergoing slopewash

Typical slope where slopewash is important

Short term weathering processes, already mentioned above in the destruction of rills, are very important in preparing slope materials for transport by flowing water. Cycles of wetting and drying, freeze–thaw, and animal disturbance are all examples of processes which can loosen a soil surface between storms. The result is a layer of highly mobile sediment which can be removed very easily during the next storm. After a period of very high initial sediment loss from a slope, therefore, the rate of erosion decreases markedly as the more resistant, intact soil beneath is exposed. This sequence may also be observed on an annual basis, in that the first few storms after a seasonal dry period will cause higher sediment loss than subsequent ones.

5.2.5 Subsurface mechanical erosion on hillslopes

Eluviation

Water infiltrating into the soil during rainfall can sometimes carry particles of sediment vertically through the soil: a process known as *eluviation*. If movement takes place between soil grains, only very small particles indeed can be transported (*Figure 5.11*). For example, even in a coarse, clean, sandy soil, one could only expect silt-size particles to be moved between the soil grains. In finer soils, or those with a wide range of grain sizes, it is really only possible for clay-size particles to be moved. This is often referred to as clay translocation. Larger particles could be moved, and more transport take place, through

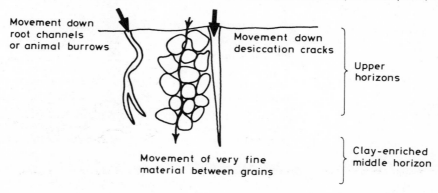

Movement down root channels or animal burrows

Movement down desiccation cracks

Upper horizons

Clay-enriched middle horizon

Movement of very fine material between grains

Figure 5.11 Vertical sediment transport in soils

secondary pore-spaces, such as desiccation cracks and root channels. Evidence for eluviation is to be found in the frequent occurrence of higher clay contents in the middle horizon of soils than in the upper or lower horizons. Clay coatings are also to be seen along cracks and root channels in some soils. It has therefore been suggested that these factors are due to the vertical washing of fine particles down the profile. Other explanations are possible however. One which seems likely is that the enriched clay content is the result of greater chemical weathering in the wetter, middle horizon of soils.

Assuming that eluviation does take place in some soils, and it probably does, the question arises of whether or not it has any significance in slope development. For it to be important here, not only must there be significant vertical movement of soil particles, but lateral movement out of the profile must also occur. There are good reasons to say that such lateral transport cannot be very important. First, the rate of movement of particles in soil water draining through the pore-space is very slow indeed, and the number of particles involved is small. Second, eluviation is a self-limiting process because as particles are moved vertically through the soil to accumulate lower in the profile, they progressively block up the soil pores and restrict further movement. On these two counts eluviation will always be of minor importance in comparison to all other slope processes. From the point of view of soil development, however, it can be a very significant process. In Chapter 5.3 we shall see that a similar conclusion is reached with respect to the movement of certain chemical constituents in soil water.

Pipes and pipeflow

In certain soils, subsurface channels called *pipes*, which discharge water during and sometimes after rainfall, are to be found. There appear to be two main types of pipe: smaller diameter (5–10 cm), which are typical of humid temperate soils and larger features (up to a metre or more in diameter), which are common in, though not exclusive to, semi-arid to subhumid soils. The smaller pipe features have been found at many sites in the British Isles and all the evidence suggests that they are widespread. They appear to form in soils which have a high permeability layer on top of a much lower permeability horizon. Frequently this break occurs at the junction between the organic soil and the mineral soil below. These pipes form poorly integrated, discontinuous systems, whose trend is largely downslope. Their origin is still under discussion. It has been suggested that they develop from drying cracks in the surface layer, but they may also be due to animal burrowing.

The larger pipes, which predominantly occur in semi-arid regions, are often large enough for human access, and are superficially similar to limestone caves in that they have an input depression (the 'sink') and an opening at the downstream end (the 'resurgence'). They also form integrated networks. Drying and cracking of the surface soil layer provides well-defined water inputs to the soil, and when the infiltrated water reaches a lower permeability soil horizon, it is diverted laterally downslope. The outlet is in a free face, such as a channel bank, and erosion by the outflowing water at this point initiates passage formation, which progressively retreats into the soil. Ths process is certainly aided by an easily erodible soil layer immediately above the permeability break

Hydrologically pipes may be regarded as subsurface extensions of the permanent channel system because they transmit water at open channel velocities. However, one must add the proviso that they are not always continuous systems, and are therefore not necessarily as efficient as surface streams in transmitting water. From the slope development point of view their importance is unclear, although it is certain that significant amounts of sediment are transported along them. Pipes in semi-arid regions seem to continue to develop until they become too large and collapse, thus forming channels. The smaller, humid temperate pipes do not seem to be developed in the same way; possibly because any particular pipe does not remain an active flow line for very long. As far as we can tell they do not seem to have an obvious effect on slope form, although we can assume that they aid removal of sediment by flowing water from slopes and that they contribute to random creep when they are abandoned and collapse.

5.2.6 Experiments to measure rainsplash and slopewash

Rate of rainsplash erosion
Rainsplash erosion is not easy to measure because the soil surface is always disturbed by installing the measuring apparatus. Also, the natural process is interfered with, especially if re-splashing of the soil out of the measuring apparatus is inhibited. It should also be remembered that significant rainsplash erosion only occurs on sparsely vegetated slopes,

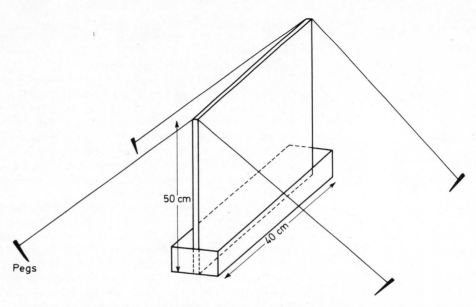

50 cm

40 cm

Pegs

Figure 5.12 A possible design for a splashboard to monitor rainsplash

which are fairly uncommon in Britain. However, if suitable slopes are to be found, the following simple piece of equipment may be used to measure rates of splash erosion.

The splashboard, illustrated in *Figure 5.12*, consists of a galvanised iron sheet with a trough fixed to both sides to catch splashed soil. It is installed with the sheet parallel to the contours of the slope and may be secured in position with guy ropes (*Figure 5.12*). After a rainstorm, the material stuck to the downslope and upslope sides of the sheet and trapped in the respective troughs is collected *separately*. Care must be taken to wash out the sediment from the troughs so that none is lost. The upslope and downslope trough and sheet samples are then dried and weighed and the soil collected from the upslope transport side is subtracted from the downward transport side to obtain a *net downslope* transport for the rainstorm.

The splashboards can be put on different slopes, different soil types or different vegetation types, to study the influence of these factors on the splash process. However, since a great many factors influence rainsplash it may be difficult to demonstrate clearly the influence of any single one. The rate of transport in any given time period is found by dividing the net weight of soil travelling downslope by the width of the splashboard across the slope. The transport rate will then be in units of grams of soil transported per centimetre width of slope (gm cm^{-1}). If the time period is one year the units become gm cm^{-1} yr^{-1}, and observations of any length can be converted to a yearly rate by dividing by the number of days of observation and multiplying by 365.

The influence of slope angle upon rainsplash erosion

Two simple experiments are described here to demonstrate the influence of slope angle on the amount of soil dislodged by rainsplash during a storm (or storms). The first allows measurements to be made, whereas the second is really for demonstration purposes only.

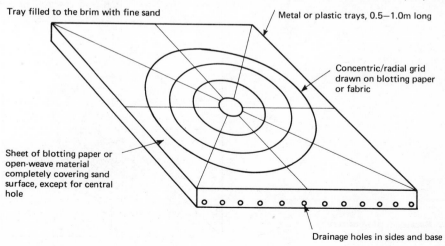

Tray filled to the brim with fine sand

Metal or plastic trays, 0.5—1.0m long

Concentric/radial grid drawn on blotting paper or fabric

Sheet of blotting paper or open-weave material completely covering sand surface, except for central hole

Drainage holes in sides and base

Figure 5.13 An experimental setup to examine the influence of various factors on rainsplash

To measure the effect of rainsplash erosion on different slope angles requires a series of test trays to be prepared, as illustrated in *Figure 5.13*. Each tray should be about 50–100 cm in length (downslope dimension) and should be drilled with drainage holes in the base and sides. Cheap plastic seedling trays of sufficient size would be suitable for the experiment, or purpose-made boxes constructed in plywood could be used. The trays are loosely filled to the brim with sand and are levelled smooth with a straight-edge. They are then covered with a sheet of blotting paper or open weave material which has a central hole of roughly 10 cm diameter, and is marked off with a series of concentric and radial lines to form a grid (*see Figure 5.13*).

Several trays are thus prepared, and stood in an exposed site outdoors propped at different slope angles, to receive a rainstorm of suitable intensity. Material exposed in the central hole will be splashed onto the sheet of paper or cloth. After exposure, the cloth or paper is dried and cut up along the lines previously drawn upon it, being careful to label the original position of the sections with respect to the central hole. It is helpful if the grid is drawn on the paper such that the area of each section is the same. If the paper

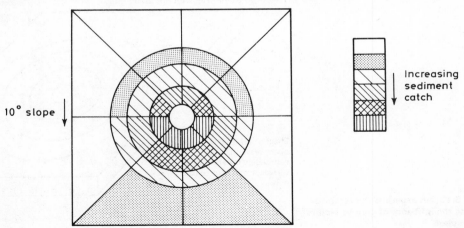

10° slope

Increasing sediment catch

Figure 5.14 Representation of results from the rainsplash experiment

134

or cloth has a reasonably consistent area density (i.e. weight/unit area) (this can be checked by weighing a number of equal area samples of the material from a similar sheet) one merely has to weigh the pieces to record the weight of sediment caught on each. If not, the sediment must carefully be washed from each segment of paper, dried, and then weighed. The results may be recorded on a diagram, as shown in *Figure 5.14*.

To demonstrate the effect of slope angle on the splash process, small trays (20–30 cm long) can be filled to the brim with sand and levelled. Several metal counters are then placed on the sand surface in each tray, and the trays are then all exposed to rainfall, propped at different slope angles. The amount of splash is roughly indicated by the height of the columns of sand protected by the metal counters. This should adequately demonstrate that the process operates and that slope angle is an important factor, but it will not allow quantitative measurements to be made.

Measuring surface water erosion

Surface water flow erosion can be measured independently of rainsplash with a sediment trap similar to that illustrated in *Figure 5.15*. A suitable sediment trap consists of a metal

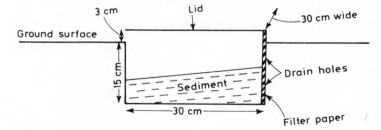

Figure 5.15 A typical sediment trap to measure slopewash

box of the dimensions illustrated in *Figure 5.15*, equipped with a lid to protect it from rainsplash. The box is open on the upslope side. It is dug into the ground so that the upslope lip is exactly flush with the ground surface, and it should be left for a few weeks to stabilise before any observations are made. Sediment is collected from the trough at regular intervals and is dried and weighed. The weight of sediment trapped shows the rate

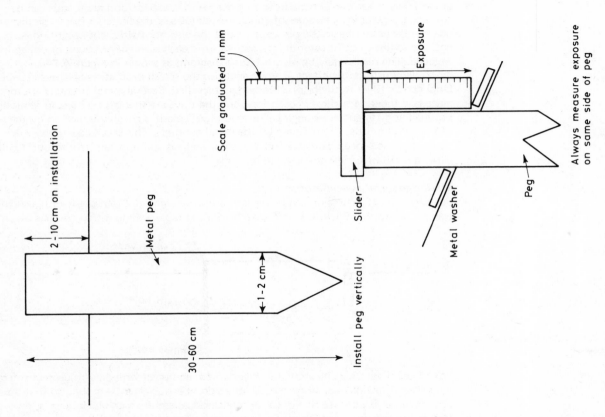

Figure 5.16 The installation and measurement of erosion pegs

of downslope sediment transport during the period between observations and should be expressed in units of gm cm slope^{-1} year^{-1}, as for rainsplash. Troughs can be installed at various distances from the crest of a slope to show downslope variations in transport rate; on slopes of different inclination to show the influence of slope angle; or on different vegetation covers. Rates of transport by flowing water are very low on well vegetated slopes, and so it is really only worth measuring this process on slopes where some bare ground occurs. This is not common on British slopes but does occur, especially under forest or scrub vegetation, on some coastal slopes, and on upland peat slopes, as well as on many cultivated slopes for long periods of the year.

Total slopewash erosion

The methods discussed above allow rainsplash and surface water flow erosion to be measured separately. A much simpler approach can be adopted to measure total slope-wash erosion, if it is not required to differentiate between the two. Metal stakes can be driven into the ground and the length of exposure measured from time to time with a simple depth gauge (*Figure 5.16*). It helps to place a metal washer over the peg when the measurements are made in order to standardise the measuring conditions, and always to measure erosion on the same side of the stake. Erosion is measured directly in terms of ground lowering, and not in terms of weight of soil loss from unit width of slope.

5.3 CHEMICAL EROSION BY WATER

In Chapter 3.2.6 we discussed some of the basic principles of chemical weathering and chemical erosion. In this section we will attempt to show how these chemical processes operate on a hillslope. In Chapter 1.4 it was shown that under natural conditions erosion and weathering are generally in balance so that the landsurface is lowered through geological time with a soil profile remaining intact on its surface. Although the particular materials forming the soil are constantly being removed by erosion, weathering of the bedrock beneath the soil is constantly adding new material to the soil so that the structure of soil over bedrock remains the same even though the material itself is being constantly replaced. For this reason it is necessary to consider weathering and consequent chemical erosion in the context of a soil profile. On most parts of the earth's surface, the soil supports plant life and plants are therefore inextricably bound up in the processes operating.

5.3.1 Hydrolysis — An important weathering process

Put very simply, weathering is the breakdown of mainly complex silicate minerals into simpler structures, notably clay minerals. The dominant weathering process is *hydrolysis*, in which the small and very reactive hydrogen ion, H^+, replaces metal ions, particularly Ca^{2+}, Mg^{2+}, Na^+ and K^+ in silicate mineral structures and also releases some silicon ions, Si^{4+}, in soluble form. These ions are highly soluble in the soil environment and are carried away by water.

In the long term, the amount of water draining over and through the soil is determined by the ratio of rainfall to evapotranspiration (*see* 5.1.1) so that the higher the ratio, the greater the chemical erosion. An excess of rainfall over evapotranspiration also encourages plant growth. Hydrolysis, and the chemical erosion which results from it, depends on an adequate supply of hydrogen ions. In all environments hydrogen ions are made available by the solution of atmospheric carbon dioxide in rainwater to form carbonic acid as:

$$H_2O + CO_2 \rightarrow H_2CO_3$$

and the ionisation of carbonic acid in solution into hydrogen ions and bicarbonate ions as:

$$H_2CO_3 \rightarrow H^+ + HCO_3^-$$

The presence of a plant cover serves in a number of ways to increase the H^+ concentration in water after it enters the soil. Living plant roots react with the surrounding soil to gain nutrients and in so doing increase the hydrogen ion concentration by secreting carbon dioxide, which ionises in solution in the manner described above. Carbon dioxide is also one of the products of organic decay in the soil and this provides an additional supply of hydrogen ions to the soil water. In this and many other more complex ways, both living and dead organic matter serve to increase the hydrogen ion concentration of soil water (which lowers the pH) and thus aids weathering and erosion. It therefore follows that

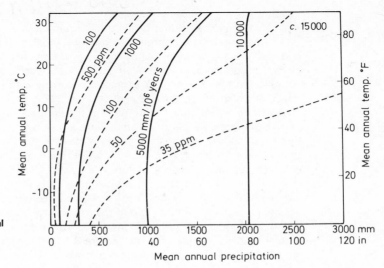

Figure 5.17 Estimated chemical removal from an igneous rock soil in differing climatic conditions (After Carson and Kirkby, 1972)

the highest rates of weathering and chemical erosion tend to be found in the areas of highest rainfall (*see Figure 5.17*).

5.3.2 Chemical loss from slopes

In calculating the amount of solutional loss from hillslopes it is necessary to subtract from the total amount observed in the water flowing from the slopes, that proportion which is derived from sources other than the bedrock. This is done by calculating a chemical budget. Rainfall contains many things in solution besides carbon dioxide; for example, calcium, magnesium, sodium, potassium, chloride, nitrogen and sulphur. The vast majority of the major anions passing through the soil system (chloride and sulphate) are derived from the atmosphere. Carbon is almost entirely derived from the biosphere so that weathering of the bedrock is mainly reflected in the loss of sodium, potassium, magnesium, calcium, and to a lesser extent silica, in drainage water. Chemical erosion is

139

measured as the difference between the amount of these weathering-derived ions in the streamflow and the amount brought in by rainfall over a given budgeting period.

A simple budgetary approach to chemical erosion considers only inputs and outputs and treats a soil covered hillslope simply as the site of reactions and transport. As has already been suggested in discussing sources of hydrogen ions, things are rather more complicated than this and the processes occurring within the soil and biomass exert important controls on the rate and nature of chemical erosion.

Leaching is the general term used for the processes of transport of solutes by water percolating through the soil. Water charged with hydrogen ions picks up cations by hydrolysis and by exchanging hydrogen for cations attached to the charged surfaces of clay minerals (*see* 3.2.2). The end result of a long period of leaching would be the complete removal of mobile cations (Ca^{2+}, Mg^{2+}, Na^+, K^+) from the soil and the relative enrichment in the soil of immobile elements such as iron, aluminium, and silica. However, although very little iron and aluminium are actually removed from the soil profile, leaching does relocate them within the soil and this is one of the important mechanisms in the development of soil horizons. Since iron and aluminium are virtually insoluble as inorganic constituents of the soil (*Figure 3.14*), there must be some way by which they are transported, if only over short distances, and just such a process is provided by organic matter in the soil.

Organic acids are released by plant roots and decaying organic matter and are soluble in the low pH environment near the soil surface which is rich in living and dead plant material. These organic acids have the ability to attach iron and aluminium atoms to their molecules and thus transport them in solution. This process is known as *chelation*.

As the water containing the organic acids moves down the soil profile it enters a higher pH environment where the organic acids are insoluble and so they are precipitated out of solution along with their attached iron and aluminium. In addition to this, the amount of iron and aluminium attached to the organic acids increases during transport down the profile and this also helps to precipitate them out of solution. The result is that iron, and to a lesser extent aluminium, are concentrated at some depth in the soil producing a less permeable 'B horizon' in the soil and in some cases even a hard pan of iron oxide, as in a podsol profile. Provided the rate of turnover of soil material is sufficiently slow to allow this process to develop fully, the result will be that less water will percolate through the B horizon to the weathering zone, and so weathering and chemical erosion are slowed down. This is a feedback mechanism which regulates the balance between weathering and erosion. Where erosion is rapid the turnover of soil-forming materials will be too fast to allow full profile development to occur and so water will percolate fully to the weathering zone at the bedrock surface. Where erosion is slow, profile development will inhibit weathering. Mechanical erosion increases with slope angle so that in general soil does not remain on steep slopes long enough for mature soil profiles to develop.

So far in considering the effects of the biosphere, we have seen examples of how plants and their decomposition products accelerate weathering, chemical erosion and soil profile development. Plants also serve to retard the loss of certain elements from the soil. Some elements are selectively absorbed by plants, that is, in greater amounts than their relative concentrations in the soil would indicate. They are then stored in the biomass and eventually returned to the soil in throughfall and leaf fall. Potassium and to a lesser extent calcium are selectively absorbed by plants and at any time a proportion of the potassium which has been released from the bedrock will be stored in plant tissues. Plants therefore slow down the rate of loss of elements from the soil by cycling them through their vascular systems. The raising of nutrients from the subsoil therefore tends to counteract leaching. If the vegetation is removed the rate of leaching of these elements increases considerably. It is important to remember that the processes which a geomorphologist sees as chemical erosion are the same processes the botanist sees as nutrient cycling.

As suggested above, the combination at any point on a slope of varying rates of weathering, chemical erosion, mechanical erosion, vegetation cycling and other related processes are reflected in the nature of the soil profile. Variations in the rates and relative importance of these processes on a hillslope profile produce a sequence of soils down the profile known as a soil *catena*. A catena is a sequence of soil profiles that occur in a regular repetition on repeated topographic sequences in a particular region. The nature of the catenary sequence varies between regions as both bedrock and climate vary. The type sequences of any area may be obscured by variations in geology. However, where the whole of a hillslope is on a single bedrock type a catena will be observed which reflects the relationships between soils and slopes for the climatic area. Careful observations of the nature of the soils in a catena are a useful guide to the processes operating. The basic principles described above apply in that where soil transport is slow, soil profiles will be well developed and this development will be inhibited by rapid soil transfer.

5.3.3 Rates of hillslope erosion by chemical processes

Most of the observations that have been made of chemical erosion refer to whole drainage basins, rather than to particular hillslopes. It is much easier to collect the necessary data for a chemical budget for a whole drainage basin than it is for a hillslope, since in the former case it can be done by sampling streamflow. It is rather more difficult in the case of individual hillslopes, though a method for collecting water draining from hillslopes is given in 5.3.4. Results for whole drainage basins will therefore give average rates for all the hillslopes in the basin but tell us nothing of variation in rate on individual slopes. This is unfortunate since it is the variations in rate downslope which lead to the development of hillslope form.

At the basin scale, rates of chemical erosion are obviously closely related to the solubility of the bedrock so that records for individual catchments must be interpreted in geological terms. However, as suggested earlier, the rate for any given bedrock should be highest where runoff is greatest. Since runoff represents the balance between evapotranspiration and precipitation, it is possible to represent the pattern in generalised terms of mean annual precipitation and mean annual temperature, where temperature is chosen as a surrogate for evapotranspiration. This relationship for a standard igneous rock is shown in *Figure 5.17*.

Variations in rates of chemical erosion on individual hillslopes are not very well understood and it is not possible at present to make any meaningful generalisations on the subject on the basis of experimental evidence. If the assumption widely used in this book is accepted, that of equilibrium between chemical weathering of the bedrock and mechanical transport of soil, then it would be expected that chemical weathering and therefore chemical erosion would be greatest on the steepest slopes where mechanical transport is most rapid. However, the situation is not as simple as this. Where mechanical transport is slow, soil profile development proceeds by the loss of mobile elements and the relative enrichment of the soil in iron, aluminium and silica, so that chemical losses from the soil are extensive (as distinct from chemical losses from the bedrock which convert bedrock into soil). Where mechanical transport is more rapid, less leached material is eroded mechanically so that the full potential loss of mobile elements from the soil does not occur. This is one aspect of hillslope development which is in need of much more field evidence.

It is possible to observe rates of solutional removal of material from hillslopes by collecting water draining from them, using the system described in 5.3.4. Chemical analysis of the water for individual ions is time-consuming and requires a great deal of expensive equipment. It is possible to measure the overall concentration of all ions in solution very easily using specific electrical conductance (conductivity). The nature of conductivity and its application to the measurement of solutes is described in the experiment in 5.3.5.

5.3.4 An experiment to collect throughflow

Water draining from the soil at the base of a slope can be collected by a simple structure made of wood and sheets of plastic (*Figure 5.18*). In this way water can be collected at any number of depths in the soil simultaneously. The best sites for constructing the collectors are where a soil is exposed in the bank of a stream. Dig a set of 'steps' in the soil, such that the tread of each step is at a level at which it is desired to collect throughflow, preferably at the base of each soil horizon. Each step should be about 1 metre long. As each step is dug, carefully retain the soil for repacking. *Figure 5.18* shows three steps, plus an additional collector for surface flow. The whole set of steps is first dug out, then a sheet of plastic 20 cm wide is laid on the first one with a slight fall along its length. At the lower end of the step a tube is fitted so that throughflow collected on the plastic can be collected in a bottle. The soil dug from this step is then carefully repacked on the plastic. The plastic sheet of the next step is then laid as for the first, only this time it must be 40 cm wide. Repeat the procedure for all steps and include one at the surface for overland flow. Finally the whole is covered with boards to keep it in place. Rates of throughflow at each level can be measured by measuring the volume of flow in a given time and hydrographs of throughflow at each level can be drawn up after storm rainfall. The dissolved solids content of the water collected at each level can be determined by the method described in 5.3.5.

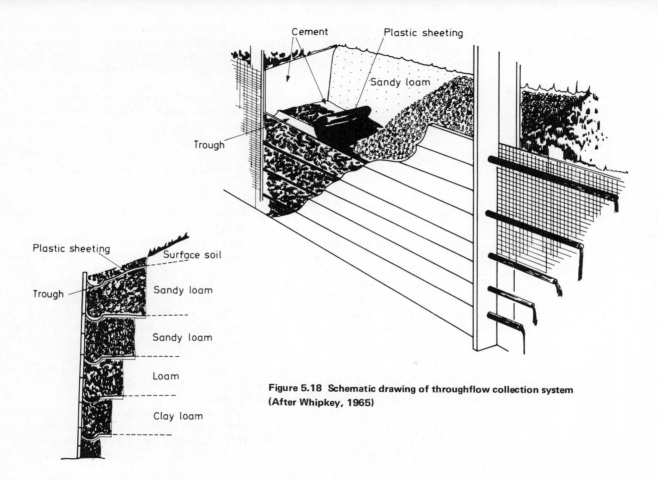

Cement

Plastic sheeting

Sandy loam

Trough

Plastic sheeting

Surface soil

Trough

Sandy loam

Sandy loam

Loam

Clay loam

Figure 5.18 Schematic drawing of throughflow collection system (After Whipkey, 1965)

5.3.5 An experiment to measure the total dissolved solid content of throughflow

Pure water is a very poor conductor of electricity. When water contains impurities in solution its ability to conduct electricity improves; thus the conductivity of a water sample is a function of its total dissolved solids concentration. For any given total dissolved solids concentration, conductivity will vary depending on the temperature of the sample. Conductivity of aqueous solutions is expressed in units of microsiemens per centimetre (μS cm^{-1}) at a specified temperature and is measured on a conductivity

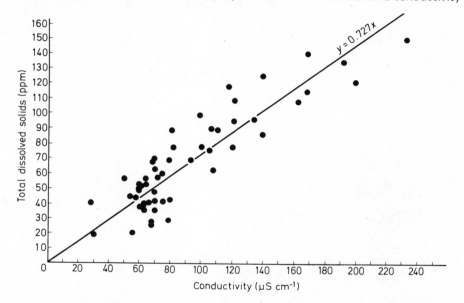

Figure 5.19 A typical calibration curve of conductivity against total dissolved solids

bridge or meter. Most conductivity meters automatically convert to a reference temperature of, say, 25 °C at the rate of 2% per degree centigrade. If the instrument being used does not automatically correct for temperature, this can be done by the operator using the formula:

$$L_{25} = L_T - 0.02\,(T - 25)\,L_T$$

where L_{25} is the conductivity at 25 °C and L_T is the conductivity of the sample at a temperature of T °C.

For each site being investigated a calibration curve can be drawn up which relates total dissolved solids concentration to conductivity. Total dissolved solids is measured by evaporating a sample of water and weighing the solid residue left behind. Total dissolved solids is then used as the dependent variable in a regression analysis with conductivity as the independent variable (*Figure 5.19*). The resulting regression equation can then be used to give the total dissolved solids concentration for any measured conductivity at the same site.

CHAPTER 6   SLOPE FORMS

6.1   MEASUREMENT OF SLOPE FORM

In studies of hillslope development three lines of attack are used:

(1)   Theoretical analyses of slope processes and their interrelationships with slope form
(2)   field measurements of processes and
(3)   field measurements of form.

We can show what slope profiles should be like in theory but we also need field evidence to relate to theory, and to form a basis for process–response models. An unfortunate feature of much early work on slope development is the almost total absence of field measurements of slope form. In this section we will present two simple methods for measuring slope profiles in the field and show how the data can be analysed and presented. We recommend strongly that measurements should be made in the field rather than from contour maps. In this section we will also show how small scale topographic features can be mapped in the field using a set of morphological symbols.

6.1.1   Field procedure

Gravity is the major force acting on hillslopes and it is at a maximum in the direction of the steepest slope. This direction can be defined on a slope by a line which crosses all the contours at right angles. Since this is the line along which processes operate, it is the *genetic hillslope profile*. One group of geomorphologists have attempted to define this line in the field by rolling a cannonball down a slope! In practice and especially in areas where contours are curved, it is sometimes difficult to define in the field. In order to overcome this difficulty a standard procedure has been adopted for locating *slope profile lines*. This also has the advantage that slope profiles measured by different people in different places are directly comparable with each other.

The method for locating a line of slope profile is shown in *Figure 6.1*. The slope profile line is a straight line in an area where the major portion of the slope is made up of rectilinear contours (i.e. contours which are nearly straight and run parallel to each other). This excludes valley heads and spur ends where the contours are strongly curved.

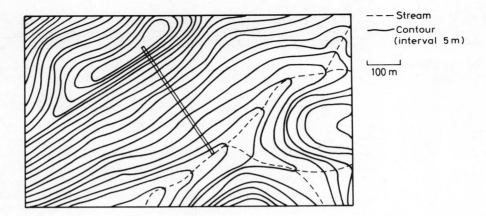

**Figure 6.1 Location of a slope profile
for measurement**

The profile line should extend from the ridge crest to the valley bottom or local base level, which would usually be a stream bed.

Measurements consist of a consecutive series of ground lengths and an angle measured relative to the horizontal plane over each length. The measured length (L) and the angle measured over it (α) together define a right-angled triangle as shown in *Figure 6.2*. The vertical fall (V) is calculated by

$$V = L \sin \alpha$$

and the horizontal equivalent (H) by

$$H = L \cos \alpha$$

The ground lengths should all be equal and as short as possible, within the constraints imposed by time available and the total length of the slope. Two instrumental techniques are commonly used, one of which requires a commercial instrument, either an *Abney*

Level or an *inclinometer*, and the other a home-made device called a *pantometer*. We recommend the latter since it is cheap, easy to build and is accurate to the nearest 0.5°, which is similar to the accuracy of a hand-held Abney Level.

The Abney Level and the inclinometer are used in conjunction with a tape measure or a piece of string of known length. Five metres is a suitable measuring length. The measuring length may be varied depending on local conditions and the purpose of the survey, but as far as possible should be kept constant on individual profiles to facilitate

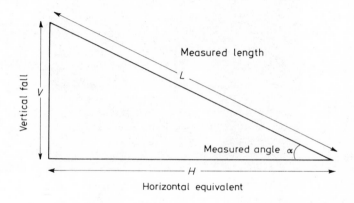

Figure 6.2 Geometry of slope measurements

subsequent analysis. It is most convenient to have two people working together such that they stand at the ends of the measured length. The person with the instrument sights it on the other, aiming at a point the same height above the ground as the observer's eye level.

A pantometer is a large wooden or aluminium parallelogram made as described in 6.1.4 (*Figure 6.3*). It is consecutively relocated down the slope profile line such that the position occupied by the front leg at one reading is occupied by the rear leg for the next. Note that it is normal, and easier, to measure slopes from the top down. Slope angles are

Figure 6.3 A slope pantometer constructed with angle-section aluminium

read from the protractor on the pantometer and the ground length is the distance between the legs (1.5 m in the example in 6.1.4).

6.1.2 Slope profile analysis

In analysing a slope profile the aim is to divide the profile into its component forms and to describe their characteristics. Variations between one slope angle and the next can be due to local irregularities or to larger scale changes along the profile. At this larger scale there are three fundamental slope forms as shown in *Figure 6.4*.

A *straight slope* is a section of a profile on which the mean angle does not change with distance. A *cliff* or *free face* is a special form of straight slope. A *convexity* is a section of profile on which slope angle increases downslope. A *concavity* is a section of a profile on which slope angle decreases downslope (*Figure 1.1*, Chapter 1.1.1).

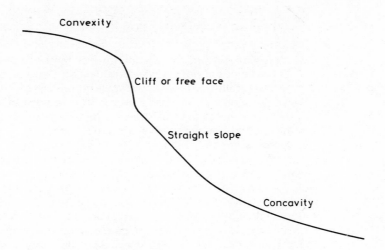

Convexity

Cliff or free face

Straight slope

Concavity

Figure 6.4 Fundamental slope forms

A simple method for subdividing a profile into these forms is shown in 6.1.5. Straight slopes are best described by the mean angle and it is also instructive to produce a histogram of the angles of straight slope (*see* Chapter 6.2). Curved slopes are described in terms of their rate of curvature in degrees per 100 metres, which is positive for convex curvature and negative for concave curvature. The formula for calculating curvature (C) over a measured length is:

$$C_b = 100 \frac{\alpha_a - \alpha_c}{2L} \qquad °/100 \text{ m}$$

where a, b and c are three consecutive slope measurements having fixed length L and angles α_a, α_b and α_c. C_b is the rate of curvature of section b. Alternatively the curvature at a point on a profile may be calculated by:

$$C = 100 \frac{\alpha_a - \alpha_b}{L} \ °/100 \ m$$

which is the curvature at the point at which measured section a and b meet.

Two methods of presenting profile data are shown in 6.1.5. A *line graph* is a plot of slope angle against distance downslope and this can be used to subdivide the profile into the component forms defined above. A straight slope appears on the line graph as a trend of points parallel to the *x*-axis (i.e. the distance axis). Convexities appear as rising trends in the line and concavities as falling trends.

A *cross-section* is a representation of the slope profile in the vertical plane. It is produced by progressively summing the vertical and horizontal equivalents (*Figure 6.2*) and plotting each point on the profile as coordinates on a graph of height (*y*-axis) against distance (*x*-axis). Some vertical exaggeration may be used where the height of the slope is small relative to its length but this should never exceed 5. Sometimes, profile data are plotted directly as sections inclined at angle α, using a ruler and protractor. It should be stressed that this is much less accurate than plotting coordinates for each point, since errors are cumulative. An example of a profile survey, together with calculated coordinates and the final plotted profile is given in 6.1.5. The method for calculating vertical exaggeration is also given in 6.1.5.

6.1.3 Morphological maps

Morphological maps (*Figure 2.4*) are used to supplement contour maps, through the use of a set of symbols to show features in greater detail than is possible on most contour maps. In order to fulfil this purpose adequately, morphological maps should be drawn at a scale of 1 : 10 000 or less. Production of these maps requires intensive field work and they are usually produced only for small areas and for special purposes. In practice they are most widely used in engineering site investigations, especially in areas which are, or have been, the site of mass failures. They are also a useful introductory exercise in geomorphology, as they help to develop an awareness of the complexities of surface topography.

Morphological maps are constructed on specially prepared base maps drawn to a scale appropriate to the study, bearing in mind the smallest feature which can be portrayed at any given scale. For example, many of the morphological symbols shown in *Figure 2.4* would be about 3.0 mm wide when drawn on a map in the field. At a scale of 1 : 10 000 a symbol 3.0 mm wide would represent a width of 30 metres on the ground so that no feature less than 30 metres across can be accurately shown on the map. The base map is prepared from aerial photograph enlargements or a small scale map if available (e.g. the Ordnance Survey 1 : 10 000 series) and should show as many fixed reference points as possible such as roads, fences, buildings, property boundaries and streams.

The basic set of symbols used in morphological mapping are shown in *Figure 2.4*. The basis of the technique lies in plotting breaks and changes in slope; sharp breaks with a solid line, more gradual changes with a broken line. V-symbols are added to the lines, pointing downslope and positioned on the side of steeper slope. This enables the distinction to be made between concave and convex changes, or breaks, in slope. Separate symbols are used for features such as cliffs and thalwegs (the thalweg is the line of the deepest channel in a stream bed) and the maps may be supplemented by slope angle measurements (*Figure 2.4*).

6.1.4 Experiment to construct and use the slope pantometer

The slope pantometer consists of two uprights linked by two cross-pieces, with the bolts connecting the cross-pieces to the uprights exactly 1.5 metres apart. A large wooden blackboard protractor is centred on one of the upper bolt holes and a reference mark is inscribed on the upper cross-piece so that the angle between the cross-piece and the uprights can be read from the protractor. A small builder's spirit level with a vertical bubble is attached to the upright on which the protractor is fitted. The reference mark is inscribed opposite zero on the protractor when the uprights are exactly vertical and the cross-pieces exactly horizontal. Angle section aluminium is the most suitable material for the frame, but wood may also be used. A typical pantometer construction is illustrated in *Figure 6.3*.

The pantometer can be successfully operated by one person though it is convenient to have two, one to read the angle and the other to record the readings. The uprights are held vertical as indicated by the spirit bubble and the slope angle is read to the nearest 0.5° from the protractor.

TABLE 6.1 DATA FOR A SLOPE PROFILE MEASURED WITH A 1.5 m PANTOMETER. THE ANGLES WERE MEASURED CONSECUTIVELY STARTING FROM THE TOP OF THE SLOPE. THESE VALUES HAVE BEEN USED TO PLOT FIGURE 6.5

Angle (degrees)	V (m)	Cum V (m)	H (m)	Cum H (m)
1.5	0.04	0.04	1.49	1.49
3.0	0.08	0.12	1.49	2.98
2.0	0.05	0.17	1.49	4.47
1.5	0.04	0.21	1.49	5.96
2.5	0.07	0.28	1.49	7.45
2.0	0.05	0.33	1.49	8.94
2.0	0.05	0.38	1.49	10.43
3.0	0.08	0.46	1.49	11.92
2.5	0.07	0.53	1.49	13.41
3.0	0.08	0.61	1.49	14.90
4.0	0.10	0.71	1.49	16.39
3.5	0.09	0.80	1.49	17.88
4.0	0.10	0.90	1.49	19.37
4.5	0.12	1.02	1.49	20.86
5.5	0.14	1.16	1.49	22.35
7.5	0.20	1.36	1.48	23.83
9.0	0.23	1.59	1.48	25.31
12.0	0.31	1.90	1.46	26.77
8.0	0.21	2.11	1.48	28.25
9.5	0.25	2.36	1.47	29.72
12.0	0.31	2.67	1.46	31.18
8.0	0.21	2.88	1.48	32.66
11.0	0.29	3.17	1.47	34.13
8.5	0.22	3.39	1.48	35.61
9.0	0.23	3.62	1.48	37.09
8.5	0.22	3.84	1.48	38.57
9.5	0.25	4.09	1.47	40.04
9.0	0.23	4.32	1.48	41.52
10.0	0.26	4.58	1.47	42.99
8.0	0.21	4.79	1.48	44.47
9.5	0.25	5.04	1.47	45.94
11.0	0.29	5.33	1.47	47.41
9.0	0.23	5.56	1.48	48.89
8.5	0.22	5.78	1.48	50.37
9.5	0.25	6.03	1.47	51.84
7.5	0.20	6.23	1.48	53.32
7.0	0.18	6.41	1.48	54.80
7.5	0.20	6.61	1.48	56.28
5.5	0.14	6.75	1.49	57.77
5.0	0.13	6.88	1.49	59.26
6.0	0.16	7.04	1.49	60.75
5.5	0.14	7.18	1.49	62.24
5.0	0.13	7.31	1.49	63.73

Table 6.1 and Figure 6.5 show some representative pantometer measurements with the calculated coordinates for the profile.

6.1.5 Representation of profile data

The data shown in *Table 6.1* were measured using a slope pantometer (*Figure 6.3*, Chapter 6.1.4). Angles were read from the protractor to the nearest 0.5°. For each angle shown in the table, the measured length, *L*, is 1.5 m. The vertical fall, *V*, and the

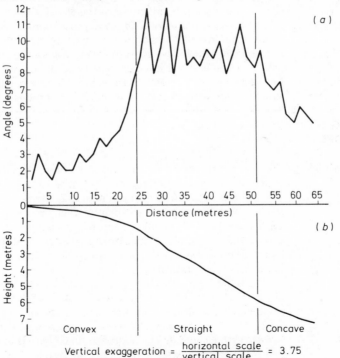

Figure 6.5a and b Line graph and cross-section of a hillslope using the data shown in *Table 6.1*

Vertical exaggeration = $\dfrac{\text{horizontal scale}}{\text{vertical scale}}$ = 3.75

horizontal equivalent H (Chapter 6.1.1, *Figure 6.2*) are also shown and each is progressively summed (cumulated) in the columns headed *Cum V* and *Cum H*.

Figure 6.5a is a line graph of the data in *Table 6.1* where angle (α, on the y-axis) is plotted against the (cumulated) distance from the top of the slope on the x-axis. *Figure 6.5b* is a cross-section of the slope plotted using the same horizontal scale as *Figure 6.5a*. The y-axis of *Figure 6.5b* is vertical fall (*Cum V* in *Table 6.1*) and the point of origin (0,0) is the top left hand corner of the graph.

The line graph, *Figure 6.5a*, shows that there is variation between successive angles due to local irregularities on the slope (slope roughness) and there are also general trends in the angles which have been shown on *Figure 6.5*. The use of the line graph and cross-section together make it easier to distinguish between local irregularities and major slope forms. The first 25 metres of the slope is a convex segment; there is then a 27 metre long straight slope with a mean angle of 9.6°; and the last 12 metres is a concave segment.

6.2 PROCESSES TENDING TO PRODUCE STRAIGHT SLOPES

6.2.1 Process–form relationships and equilibrium

Throughout this book we have taken the view that slopes are formed by the processes acting upon them and therefore it is logical to explain them in terms of those processes. This concept was introduced in Chapter 2, where different ways in which slopes could be studied were discussed. The background chemistry and mechanics to hillslope processes formed the bulk of Chapter 3, and the processes themselves were explained in Chapters 4 and 5. The remaining sections of this chapter forge the links between processes and form and will show that specific processes may produce distinctive forms, which can be observed in the landscape.

The basic descriptive tool in hillslope geomorphology, the hillslope profile, was introduced in Chapter 1.1 and discussed in some detail in Chapter 6.1. You may recall that a slope may be straight (rectilinear) or curved (concave or convex) in form. Usually slopes are composite, consisting of several curved and straight segments, such as those illustrated in *Figures 1.1* and *6.4*. The approach taken here is to look at processes generating straight, concave, and convex segments of profile respectively.

Of prime importance in relating slope form to process is the principle that form tends towards an equilibrium during the prolonged operation of a process. Thus, irrespective of its original form, as time passes a slope becomes modified to a shape which is characteristic of the process. When equilibrium is attained, the slope profile may continue to be lowered by erosion without any overall change in shape. How this may be achieved is illustrated in *Figure 6.6*. Through a period of stream downcutting into a new surface the slopes will change rapidly at first, attain equilibrium and be lowered unchanged, and then when the limit to downcutting is reached by the stream, the slopes will begin to flatten. We must

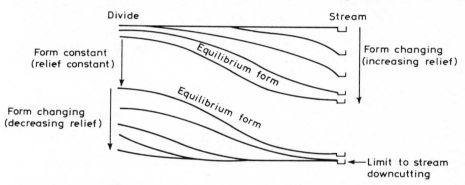

Figure 6.6 Development of an equilibrium profile

stress that *Figure 6.6* is not intended to be a general slope evolution model for all (or even any) slopes. It is merely to show geometrically how an equilibrium slope may be attained and survive in the landscape. Notice that the stream position remains fixed in *Figure 6.6*. This is an oversimplification, since it may shift laterally into or away from a slope, or it may even shift vertically upwards by aggradation, for a short time. However, in a landscape which is constantly being lowered by erosion, aggradation can only be a short-term local process. What is happening to the position of the stream is of key importance in slope development however, which will be shown for straight slopes below.

It is a basic assumption in this chapter that all slopes will tend towards equilibrium under constant conditions. However, it is more than likely that many slopes will never reach it, because of changing environmental conditions during their history. Nevertheless it is still important to be able to predict equilibrium forms for specific processes, since this allows comparison with real slopes and aids landscape interpretation.

6.2.2 Straight slopes

Straight or rectilinear slopes have a more or less constant angle throughout their length, and consequently their form can be described simply in angular terms. Very few slopes are completely straight, and even those which appear to be so usually have quite a lot of variation in angle if they are surveyed accurately. However, if the slope shows no consistent trend towards changing angle over its length it may be considered straight (*Figure 6.7*). Many slopes possess only a straight segment with curved segments above and below. But if this is a consistent occurrence in an area and forms a conspicuous landscape feature, it is still worthy of explanation.

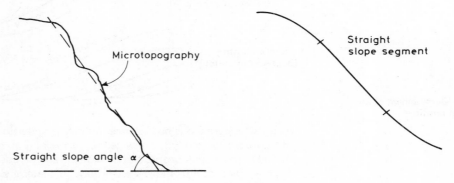

Figure 6.7 Straight hillslopes

  A straight slope may develop in a number of ways, illustrated in *Figure 6.8*. First, it may result from progressive removal of irregularities, by flattening of steeper sections and steepening of flatter sections by erosion and deposition. It may develop subsequently by *parallel retreat, hinged decline*, or *replacement* (*Figure 6.8*). How a straight slope develops depends on many factors, including weathering of the slope sediments, lateral

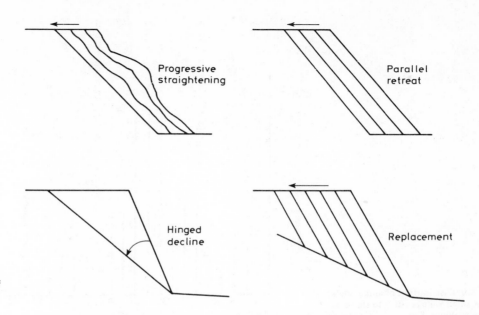

Figure 6.8 Development and retreat of straight slopes

or vertical stream erosion (controlling removal of sediment from the slopes), and of course the slope processes themselves.

6.2.3 Mean, maximum, and characteristic straight slope angles

If many slope profiles are surveyed in a region, or in a single valley where straight slopes occur, a range of straight slope angles will be encountered. This can be illustrated by a histogram of straight slope angles for some small valleys cut into slaty bedrock in County Wicklow (Eire) (*Figure 6.9*).

Straight slope segments were identified on over sixty profiles surveyed with a pantometer (Chapter 6.1.4), and plotted to give the histogram in *Figure 6.9*. The histogram is corrected for different lengths of the straight segment from slope to slope. Straight

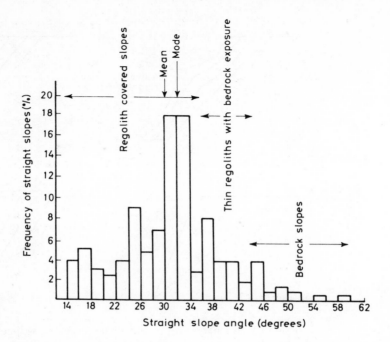

Figure 6.9 Distribution of slope angles in some small valleys cut into slates in County Wicklow, Eire

slopes range from 14–60°, though the distribution is strongly skewed towards the lower angles. *Maximum* straight slopes in this area are therefore up to 60°, though these are always bedrock and quite short. The upper limit for regolith covered slopes is about 44°, though even these slopes show thin soils and frequent bedrock exposure. The histogram shows a marked modal peak at 30–34°, and the calculated mean for the distribution is 30°. The 30–34° straight slope clearly forms a major landscape feature (over 30% of all the straight slopes fall in this range), has a causal link with the slope forming processes, and may justifiably be called a *characteristic* slope angle. There is a sharp cutoff in the distribution at 14°, with no straight slopes occurring at lower angles. This may be some kind of **threshold limit**, for the straight slope forming processes operating in this area. That is, processes responsible for the production of straight slopes do not operate on slopes gentler than 14° in this area. The distribution of straight slope angles therefore tells us much about the processes forming the slopes.

In the example used here it has been possible to define limits for the process and to indicate the typical straight slope in this area. A word of caution is in order here however. Only 32% of the straight slopes lie in the modal peak. It follows that 68% do not. Accordingly, explanation of the process in terms of limiting and characteristic angles is only a starting point for discussion, and we must try to understand the whole distribution. No clue has yet been given to the identity of the process forming straight slopes. This is rectified below.

6.2.4 Rapid mass movements and straight slopes

Chapter 4.2 demonstrated that landsliding in cohesionless soils is governed only by the slope angle. That is, sliding commences at some critical angle for the process and ceases at a lower critical angle. Imagine a river cutting laterally into the base of a slope composed of cohesionless material, such as sand or gravel. As undercutting proceeds, the slope angle will increase until the critical angle is reached and sliding commences. If basal attack continues the whole slope will eventually be replaced by a straight slope, governed by the critical angle for the landsliding process (*Figure 6.10*).

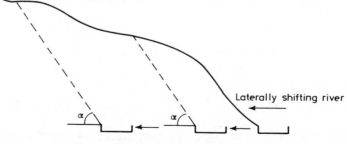

Figure 6.10 Replacement of a river bluff by landsliding to produce a straight slope

Successive positions of the river channel

You should re-read the experiment described in 4.2.6 at this point, because important lessons can be learned about the relationship between form and process on slopes controlled by landslides (rapid mass movements). First, notice that the slopes produced by the simple tilting box experiment are always straight. This is entirely independent of initial form. If you make a very irregular initial surface on the sand in the box it is progressively replaced by a straight slope as the box is tilted. Secondly, there are two critical angles which govern sliding. An upper angle controls the initiation of the slide

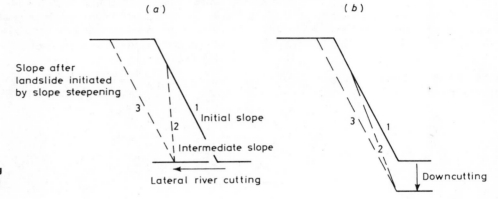

(a)

Slope after
landslide initiated
by slope steepening

3

1
Initial slope

2

Intermediate slope

Lateral river cutting

(b)

3
1

2

Downcutting

Figure 6.11a and b Cycles of landsliding on river bluffs due to lateral and vertical stream incision

while a lower one controls its coming to rest. This is one reason why there can be a range of slope angles on slopes forming on the same material; it depends on how recently a slide took place (*Figure 6.11*). Thirdly, the angle of the slope is related to the strength of the material on the slope and the water conditions within it. For a dry sand the upper angle (upper angle of repose, *see* Chapter 4.2.1) is approximately equal to the angle of internal shearing resistance, which is a precise measure of frictional strength (Chapter 4.2.1). If water fills the pore-spaces of the soil it may exert a pressure, which tends to force the grains apart and reduce friction between them. Hence, the critical stable angle is reduced. It is important to refer to 3.1.6, 3.1.7, 4.2.1, and 4.2.6 to understand these points fully.

To conclude, straight slopes or parts of slopes are to be expected where rapid mass movements operate as a major slope forming process. They will occur whenever the critical stable slope angle for the soil is exceeded in cohesionless materials, and are therefore usually common in areas of relatively rapid vertical or lateral erosion. If the soil possesses some cohesion the picture is a little more complex, and this situation is discussed in 6.2.8. Three case histories of slope studies in areas of straight slopes are discussed below.

6.2.5 Slopes due to rapid mass movements in Exmoor and Pennine valleys (Carson and Petley, 1970)

Hillslopes forming on Millstone Grit in some Pennine valleys and Devonian slates and shales in Exmoor valleys were examined in this study. The rocks break down to give a loose mantle of fine sandy soil with incorporated rocky debris, up to 1.5 metres thick. This material is practically cohesionless and so this case study provides an illustration of cohesionless slopes controlled by mass movement, where slope angle is the only control of the sliding process. The slopes are steep in both areas due to rapid vertical stream incision, and consequently landslides are fairly common on the valley sides. They usually occur after prolonged rainfall, which shows that water is an important factor in the stability of the slopes. In both areas the landslides are confined to the regolith and do not influence the stronger bedrock below. Accordingly, bedrock slopes will be reduced by weathering in these areas to an angle at which regolith can accumulate on them and subsequently slope angles will be related to the stability of the regolith. This is an important point; the slope angles here (and in many other areas of steep slopes) are controlled by the regolith properties and *not* those of the bedrock.

Carson and Petley set out to demonstrate the relationship between landsliding and slope form and angle in these two areas of differing bedrock. They started by surveying profiles on the slopes using tape and Abney level traverses (Chapter 6.1). Most of the slopes showed an upper convexity and sometimes a lower concavity near to the stream, but the middle section of most slopes was dominated by a conspicuous straight slope segment. The distribution of straight slopes in both areas showed distinct grouping. For the Pennine Millstone Grit slopes, three groups occurred around $20°$, $27°$, and $33°$, while for the Exmoor shale and slate slopes two groups were discernible at $33°$ and $25°$ (*Figure 6.12*). Intermediate straight slopes between the groups are not very common,

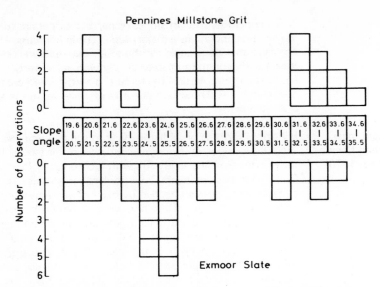

Figure 6.12 Straight slope angles on some Pennine and Exmoor slopes (After Carson and Petley, 1970)

which suggests that slopes change from one group to the next by replacement and not hinged decline.

Having established the existence of straight slopes with modal or characteristic grouping, Carson and Petley were able to study the relationships between the strength and characteristics of the regolith and the slope angle distribution. The 33° slope group occurred at the top of the valley sides and consisted of a coarse scree slope. This was interpreted as an accumulation of boulders, fallen from an original bedrock slope, and resting at its lower angle of repose (Chapter 4.2.1). Because scree is so coarse there is little chance of the pore-spaces becoming filled with water and so porewater pressures are not important in controlling the maximum stable angle.

The 21° and 25–27° slope segments were mantled with a sandy regolith with variable amounts of gravel and boulders in the matrix. Unlike scree, this type of regolith can become filled with water after prolonged rainfall. A common situation is for water entering the soil to be diverted laterally at the soil—bedrock interface and to flow parallel to the ground surface down the slope. Under very wet conditions, the whole soil may be filled to the ground surface with flowing water. Carson and Petley measured the angle of internal shearing resistance of the soils and found that the modal slope angles were close to those expected with parallel groundwater flow, as described above.

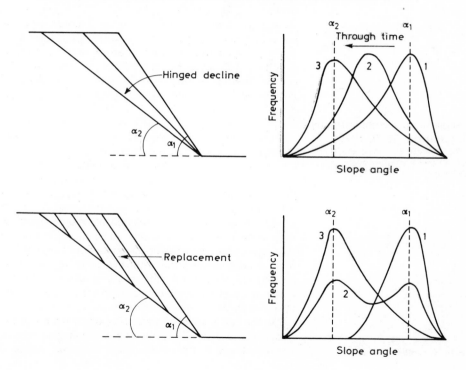

Figure 6.13 The influence of slope decline method on slope angle distribution

The two modal groupings of slope angle (25–27° and 21°) occurred on regoliths of different states of weathering and different frictional strength, clearly establishing the link between regolith properties and slope form. Further, because the three modal groups of slope angle were distinct we may infer that intermediate straight slope angles are rare and that progress from one to the other through time (by regolith weathering) is by replacement. *Figure 6.13* shows the development of the slope angle histogram by hinged decline and replacement. Slow hinged decline will result in a single mode of slope angles through time whose value completely declines, whereas replacement results in two modes with shifting emphasis from one to the other through time.

This important study by Carson and Petley, which has only received very scant attention here, demonstrates important issues. First, straight slopes in cohesionless regoliths can be shown to be due to landsliding, and secondly their inclination is controlled by the properties of the regolith which are in turn related to weathering.

6.2.6 Scree slope accumulation by rockfall (Statham, 1973, 1976)

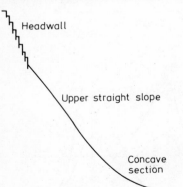

Figure 6.14 A typical scree slope

Cliff faces made of hard rock disintegrate by weathering to yield loose blocks which fall and accumulate at the base of the slope as *scree slopes*. Scree slopes are very prominent features of mountain slopes and usually form the first stage of development from a steep, bare rockslope to a regolith-covered hillslope. They usually have a straight upper segment and a concave lower slope segment, whose relative proportions are variable (*Figure 6.14*). Statham proposed that the balance between the length of straight slope and concavity was controlled by the mechanics of the rockfall process.

On a scree slope developing beneath a cliff, a sequence of profile form through time may be envisaged. Initial impact velocities of the falling boulders will be high and, after a small accumulation has occurred, downslope travel velocities on the scree will also be high. A large range of downslope travel distances will exist due to boulders falling from different heights on the cliff as well as to random factors of impact on the slope. This range of travel distances is responsible for the concave section of slope, which initially occupies the whole length of scree.

As the headwall is progressively buried the range of input velocities is diminished and this, coupled with increasing scree slope length, ensures that more and more boulders are brought to rest higher on the scree. The basal concavity decreases in overall importance and the top of the scree tends to a uniform gradient (straight slope), which approaches the lower angle of repose (Chapter 4.2.1) for the boulder accumulation. When the height of fall of the boulders approaches zero and their impact velocity approaches zero, the slope tends towards a straight slope at the lower angle of repose throughout its whole length. In the latter stages of accumulation, slides of the surface layer of scree become important in moving material downslope. This sequence of slope development is illustrated in *Figure 6.15*.

Two approaches were followed to test the validity of this theoretical argument on the development of scree slope form. First, mechanical simulations of scree slope accumulation were carried out in the laboratory by building scree slopes in a frame. Approximately 1 cm diameter gravel was dropped slowly into a frame (1 m high by 10 cm wide) from a constant height of fall, and the form was recorded at intervals during the test. A great number of simulations were performed, but all showed the same trend of slope development, which is illustrated by the example in *Figure 6.15*. The similarity between the model slopes and the theoretical discussions above is quite apparent.

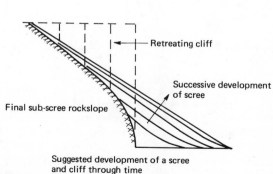

Retreating cliff

Successive development of scree

Final sub-scree rockslope

Suggested development of a scree and cliff through time

Figure 6.15 Hypothetical scree slope development and a laboratory model test compared

Input height

Height (cm)

Length (cm)

Laboratory model of screeslope accumulation, using

The second approach was to study real scree slopes in Cader Idris (Central Wales) and the Cuillin Hills (Isle of Skye, Scotland). Profiles were surveyed on many scree slopes to relate form to height of fall. First, all profiles measured on the screes showed some degree of basal concavity and an upper straight section of slope. For a group of 15 slopes in the Cuillin Hills (the only ones where headwall height could be determined with any accuracy) a relationship between height of cliff and length of concavity could be demonstrated. *Figure 6.16* shows this by a plot of a dimensionless measure of headwall height against

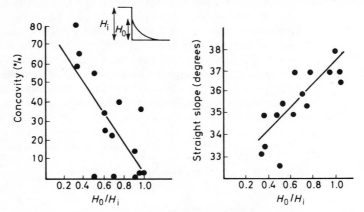

Figure 6.16 Relationships between proportional height of fall and slope form on scree slopes in the Cuillin Hills, Skye

percentage of scree occupied by a concave slope. It is apparent that concavity length decreases as the proportional height of fall of particles decreases. It was also found that the angle of the upper straight segment was related to height of fall (*Figure 6.16*). As height of fall decreases, straight slope angle increases until it approaches the lower angle of repose for the material at zero height of fall. This is due to the impact energy of boulders dislodging others on the slope at higher heights of fall, tending to move material downslope and reduce the straight slope angle.

Only a very simple scree slope system, affected by rockfall alone, was studied in this work. Many other processes operate on screes in different environments. Also the model of scree accumulation was itself a very simple one and it should be possible to improve it considerably with further research. Nevertheless, the study provides a simple example of how form might relate to the process through time; a process which seems to lead sequentially towards a straight slope.

6.2.7 Weathering versus slope recession and landsliding in cohesive materials (Hutchinson, 1975)

Hillslopes are affected by three broad groups of processes. First, weathering influences the material on the slope and is constantly changing its character. Second, hillslope processes move weathered and unweathered material to the base of the slope. Third, basal processes take material away from the slope (erosion), resulting in recession of the slope itself. Hutchinson set out to show the importance of the balance between rate of weathering and slope recession in determining the nature of the landsliding process and slope form on London Clay cliffs. In this case study we shall be looking at landslides on cohesive materials and we will see that, although some contrast with cohesionless soils can be shown, in many instances straight slopes are again to be expected from the process.

Three types of London Clay cliffs were distinguished by Hutchinson in the region of the Thames Estuary: Type 1 — rate of weathering balanced by rate of erosion at the base of the cliff; Type 2 — rate of erosion at the cliff base greater than the rate of weathering; Type 3 — rate of erosion at the cliff base equal to zero (no longer attacked by the sea).

Type 1 cliffs
On London Clay cliffs subject to moderate rates of recession (cliff top receding at between 0.2 and 0.8 m yr^{-1}) the advance of weathering into the exposed clay can keep pace with the loss of material from the slope. Thus, although London Clay is a strongly cohesive material these slopes are always covered with a mantle of cohesionless, weathered clay which has been produced by frost action, absorption of rainwater, wetting and drying cycles, and so on. This material typically consists of lumps of hard clay in a very wet, muddy matrix. Since the regolith effectively controls landsliding on a slope, it is the characteristics of this material which determine slope angle and form. Type 1 cliffs are therefore occupied by shallow landslides or *mudslides* in the wet clay debris. Material is supplied to the mudslides from the top of the cliff by small slides and is removed in equal

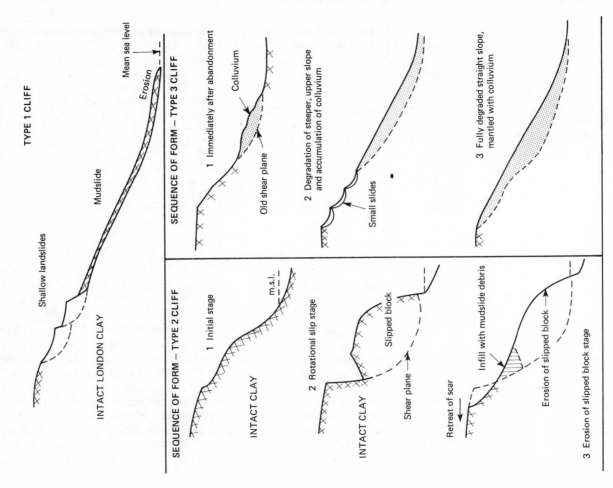

TYPE 1 CLIFF

Mean sea level

Erosion

Mudslide

Shallow landslides

INTACT LONDON CLAY

SEQUENCE OF FORM – TYPE 3 CLIFF

1 Immediately after abandonment

Colluvium

Old shear plane

2 Degradation of steeper, upper slope and accumulation of colluvium

Small slides

3 Fully degraded straight slope, mantled with colluvium

SEQUENCE OF FORM – TYPE 2 CLIFF

1 Initial stage

m.s.l.

INTACT CLAY

2 Rotational slip stage

Slipped block

INTACT CLAY

Shear plane

Retreat of scar

Infill with mudslide debris

Erosion of slipped block

3 Erosion of slipped block stage

Figure 6.17 Development of slope form on London Clay Cliffs (After Hutchinson, 1975)

amounts by the sea at the base. Usually the cliffs are entirely mantled by mudslides, and the sea never erodes intact clay (*Figure 6.17*).

Type 2 cliffs

Where rates of erosion are very fast in the London Clay (1–3 m yr^{-1} recession) the weathering processes cannot keep pace. The best example of this type of cliff in the London Clay is now the seaward face of the Isle of Sheppey (Kent). Here, marine erosion attacks the base of the cliff and steepens it until a large rotational slip takes place in the unweathered clay. After the slide, a sequence of slope development takes place while the slipped block is eroded away by the sea (*Figure 6.17*). While the slipped block remains the cliff top is protected from the direct influence of marine attack, but it still recedes due to small slips on the very steep backscar of the major rotational slip. Eventually the slipped block is completely removed, exposing a fresh, intact clay cliff, on which large scale sliding takes place again.

Type 3 cliffs

Abandoned London Clay cliffs and inland slopes do not suffer basal erosion and so weathering assumes a dominant role on the slopes. The sequence of profiles after abandonment is illustrated in *Figure 6.17*. The steep unweathered cliff degrades, rapidly at first, by rotational slips until a mantle of slipped and weathered debris progressively covers the whole slope to the angle of ultimate stability for the weathered clay.

6.3 PROCESSES TENDING TO PRODUCE CURVED PROFILES

Despite the fact that straight and curved sections of slope profiles exist together in the landscape, in harmony with each other and usually linked by smooth gradations, the dominant processes tending to produce them are very different. The theoretical basis of slope curvature is quite distinct from the theory of straight slopes as described in Chapter 6.2. Straight slope processes are controlled by threshold conditions, such that when the threshold is reached, the slope adjusts rapidly. By contrast, on curved slopes there is usually a slow evolution towards a characteristic form involving a variety of processes.

The processes tending to produce curved slopes are transport-limited whereas those tending to produce straight slopes are weathering-limited (Chapter 1.4) and this is a fundamental and important difference between the two types of slopes. On transport-limited slopes, development is limited by the transporting capacity of the processes and the rate of weathering is reduced below its potential value by the development of a stable soil profile.

In an area where stream downcutting is relatively rapid, mass failures and therefore straight slopes will be dominant. It is only when active downcutting has ceased, or is very slow, that slope angles will become adjusted to the angle of stability of the weathered mantle and the transport-limited processes which are responsible for profile curvature will become dominant. When this stage is reached the form will slowly evolve dominant curvatures, and the landscape will be lowered with little further change in form (Chapter 6.2.1) as there are no threshold angles for transport-limited processes. In humid areas the form of the curved equilibrium profile will be convex–concave while in more arid regions the convexity will not be very well developed and the profile will be markedly concave (Chapter 6.4.3).

Because of their slow rate of operation the transport-limited processes on curved slopes are relatively difficult to measure. You will notice that in the case studies on curved slopes in 6.3.2, 6.3.4 and 6.3.5 we have used a laboratory simulation to illustrate slope convexity, and the two studies illustrating concavity were carried out in areas of abnormally rapid activity.

6.3.1 Convex slopes

Convexities commonly occur on the upper parts of slopes near the drainage divide and the terms *upper convexity* and *summit convexity* are commonly used to describe them. Convexities may also occur at the base of slopes where local stream undercutting occurs.

The two most important processes responsible for summit convexities are soil creep (Chapter 4.4) and rainsplash (Chapter 5.2.3). Convex summits may also develop by cambering of resistant strata overlying a slope composed of less resistant rocks. The basic principles of the development of convex slopes are the same for both creep and rainsplash. Rainsplash will only be effective in sparsely vegetated areas (arid and semi-arid climates) while creep will be most effective in areas where soils are well developed (humid climates, as discussed in Chapter 6.4.3).

The principles applying to the development of summit convexities were first stated by G.K. Gilbert in 1909. In *Figure 6.18* the two parallel lines represent the position of the ground surface at successive times. Assuming uniform soil depth and no change of soil depth through time at any one point, the volume of soil between A and B will have passed point B and the volume between A and C will have passed point C for this amount of lowering to have occurred. The volume of soil passing progressive points downslope therefore increases with distance from the divide. If the transport rates for the processes are

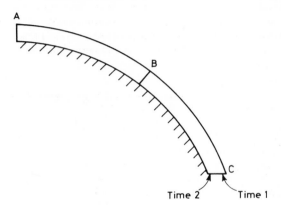

Figure 6.18 Equilibrium transport on a convex slope

proportional to slope angle then clearly the slope angle must also increase with distance from the divide to transport the sediment from the slope. Hence convex slope is the equilibrium form.

Since both creep and rainsplash act relatively slowly they can only be dominant, and therefore the major determinants of form, where slope angles are below the threshold limit for mass failure, and where wash is limited. Wash is unimportant near the divide, where the contributing area is very small, and also on soils with high infiltration capacity. The following section describes a laboratory simulation of rainsplash. The basic principles, though not the mechanisms of the process, apply equally to soil creep.

6.3.2 Rainsplash and the convexity of divides (Mosley, 1973)

This study is an example of the use of laboratory simulation in slope processes as discussed in 2.3.3. Mosley demonstrates the potential erosive power of rainsplash by pointing out that 75 cm of rain in one year over an area of one square mile has the power of about 10 000 tonnes of TNT. The principles of rainsplash erosion were described in 5.2.3. Mosley has shown experimentally that the total weight of soil splashed into air increases with slope angle. Upslope splash is at a maximum on low slope angles, when total splash is at a minimum. As the slope angle approaches $25°$, downslope splash approaches 100%. He then went on to show how this affected slope form. A wooden trough was filled with sand having a uniform distribution of grain sizes from 1.0 to 0.1 mm. This trough was then treated with artificial rainfall at an intensity of 50 mm per hour and its surface profile surveyed periodically.

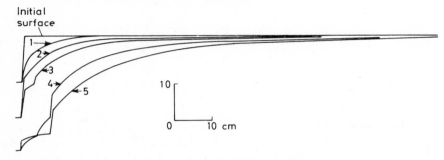

Figure 6.19 Model slope profiles developed under rainsplash

Five successive profiles are shown in *Figure 6.19* and these clearly show the convexity which developed and the way in which it extended at the expense of both the original horizontal surface and vertical face. As the convexity became steeper at its lower end slumping occurred, causing the irregularities in the profile shown in *Figure 6.19*. Mosley also observed that when the slope angle became steeper than $25°$ particles were not ejected from points of impact to any great extent and most movement was by rolling and sliding of grains dislodged by raindrop impact. For the material used in the experiment, $25°$ appears to represent a threshold angle at which rapid failures begin.

Mosley's study is limited by the fact that it was carried out in the controlled conditions of a laboratory but it demonstrates beyond doubt that where raindrop impact is the dominant processes, a convex slope will develop.

6.3.3   Concave slopes

Concave slopes may be equilibrium forms, slopes of transportation, or depositional slopes, depending on local conditions. One essential feature for the development of a concavity is that rapid downcutting is not occurring at the base of the slope. In almost all cases a concave slope is an expression of the typical hydraulic curve apparent in river long-profiles. The major exception to this is the case of scree slopes discussed in 6.2.6.

Where surface wash is the dominant process, discharge increases downslope (since the contributing area becomes larger) and therefore the transporting ability of the flow also increases. As the depth of flow increases, velocity can be maintained on lower slope angles. Particle size decreases downslope and any given flow can carry a greater load of fine material than coarse material. These factors combine to produce an excess transporting capacity in the flow so that slope can be reduced while the power to transport load is maintained. This is identical to the case of stream long-profiles and the end result is a concavity. This reasoning applies equally to transportation slopes and to equilibrium slopes where wash is dominant. Very similar reasoning applies to depositional slopes. In a depositional environment the coarser particles will be deposited first and the finest particles transported furthest.

An important factor in the production of surface wash, and therefore concave slopes, is the infiltration rate of the soil. Concave wash slopes are best developed in semi-arid areas where infiltration rates are low and rainfall intensities are high. In humid areas solution is considered to be important in the development of concave slopes, though the precise mechanisms have yet to be adequately studied.

The steepness of the concavity is closely related to the grain size of the soils, which in turn is related to bedrock type. The steepest concavities develop in coarse-grained materials and slopes of moderate concavity form on fine grained soils. In both cases the concavity is less strongly developed where a vegetation cover is present.

6.3.4   Wash erosion in Western Colorado (Schumm, 1964)

Rates of soil wash can be measured directly using erosion pins. A pin is driven vertically into the soil and changes in the distance from the top of the pin to the soil surface indicates the amount of erosion or deposition that has occurred between measurements

(Chapter 5.2.6). Schumm established several rows of pins on unvegetated slopes in Western Colorado.

Measurements of pin exposure were made on nine occasions over a four year period. No significant relationship was found between average pin exposure and slope angle during this period. However, when pin exposure is tabulated by slope segments (*Table 6.2*), erosion on the upper convex and straight segments of the profiles is ten times

TABLE 6.2   PIN EXPOSURE ON SLOPE SEGMENTS IN WESTERN COLORADO

	Slope segments		
	Convex	Straight	Concave
Average pin exposure (mm)	6.7	7.6	0.6
Average slope (degrees)	19.5	35.0	24.5
Number of pins	46	15	25

faster on average than on the concavities at the slope base. Schumm concluded that slope angle is not the factor controlling erosion rate for 'on slope segments of equal inclination, significant erosion is occurring near the slope crest, but only minor erosion or deposition at the base'. The lower concavity is not a simple depositional feature here as the form is cut in bedrock. Rather it is a transportation slope where most of the energy available is used in transporting material derived from the slope segments above.

6.3.5   Debris flows on vegetated screes in the Black Mountains, Dyfed (Statham, 1976)

In this study, debris flows were monitored in gullies on stable scree slopes and it provides us with examples of alternative ways of measuring erosion. Debris accumulates in the gullies by erosion of the gully sides and when sufficient material has accumulated debris flows occur, triggered by large storms which produce high porewater pressures in the accumulated debris. The supply of sediment from the gully walls was measured using sediment traps. These are metal boxes, set flush with ground level, which collect material eroded from a small section of gully wall (Chapter 5.2.6). Total sediment yield is calculated by applying the trap rate to the total length of walls in the gully. A second measure of total sediment yield from a gully can be calculated by measuring the volume

TABLE 6.3   MEASUREMENTS OF EROSION BY DEBRIS FLOWS IN GULLY B AND ASSOCIATED CALCULATIONS

Erosion rates on Gully B

	Eastern side		Western side		Units
	1	**2**	**3**	**4**	
Trap sediment yield	0.0128	0.0184	0.0039	0.0028	$m^3\ m^{-2}\ yr^{-1}$
Mean sediment yield	0.0156		0.0033		$m^3\ m^{-2}\ yr^{-1}$
Gully side area	435.6		497.3		m^2
Sediment yield	6.795		1.641		$m^2\ yr^{-1}$
Total		8.436			$m^2\ yr^{-1}$

Volume of debris removed from gullies in observation year (calculated from volume of deposits)

Gully	A	B	C
Volume (m^3)	11.5	9.8	8.3

Volume of gully B $= 5400\ m^3$

Estimate of gully age $= \dfrac{5400\ m^3}{8.436\ m^3\ yr^{-1}} = 640\ yr$

Or $= \dfrac{5400\ m^3}{9.8\ m^3\ yr^{-1}} = 550\ yr$

of material deposited below the gully after a flow. The long term erosion rate of a gully can be determined by measuring the volume of the gully which is the total amount of material removed during its formation. The results of these measurements and calculations are shown in *Table 6.3.*

If the observation year is accepted as representative of average years then the length of time over which the gullies have formed can be calculated by dividing gully volume by

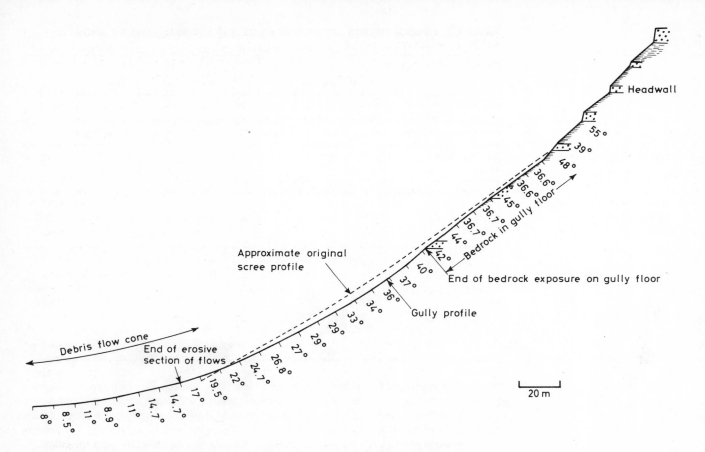

Figure 6.20 Profile of a typical gully-flow cone system in the Black Mountains (After Statham, 1976)

sediment yield in the observation year. This indicates that the gullies are between 500 and 700 years old, subject to the above assumption of course. Statham suggests that gully erosion may have been initiated by some environmental change, for example the introduction of sheep grazing in the area causing damage to the vegetation.

Of particular interest here is the fact that debris flows are converting the original relatively straight scree profile into a long concave profile down the gully axis, as shown in *Figure 6.20*. Notice that, except for that part of the gully having a bedrock floor, the concave curvature is continuous from the eroding section to the area of deposition.

6.4 SLOPE FORM IN DIFFERENT CLIMATES

6.4.1 Environmental influence and the concept of climatic geomorphology

Landforms developing in an area to some extent reflect all the environmental influences present. *Figure 6.21* is intended to show some of the important environmental controls on hillslopes and some of the ways in which they influence each other, without being in any way a complete summary of all the interrelationships.

Geology exerts important controls through rocktype and structure. Clearly, harder and more resistant rocks will tend to form areas of higher ground. Harder rocks too, show their presence on specific slopes. For example, beds of harder rock frequently form steeper sections on hillslopes. Structure, especially the dip and strike of rocks, controls the orientation of many major topographic features. The characteristic scarp and vale topography of south-east England, with more resistant strata forming the scarps, is a very good example of this major control. Structure can also influence specific slopes. For example, you may recall from 4.2.2 that the style and size of rapid mass movements (slides) on hard rocks is strongly controlled by the orientation and spacing of joints and bedding planes in the rock.

Another important control on hillslopes and hillslope processes is base level. Base level is the level to which streams are 'trying' to erode (usually sea level, but it may be controlled locally by another level, such as a lake), and its position relative to the streams influences the rate of valley lowering. In turn, this controls the steepness of the valley slopes and the rate at which weathered material is transported away from them. Streams

179

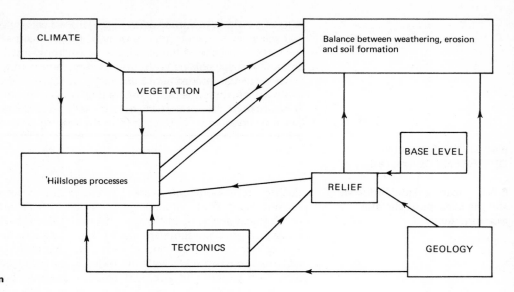

which are eroding rapidly in the vertical direction will have steep-sided valleys, and weathered material will be transported away quickly from the slopes. Naturally this has an important effect on the development of soil on slopes. As erosion rate increases, soils become thinner because supply of weathered material does not increase at the same rate to keep pace. The limit to this is naturally a bare rock slope, where transport from the slope exceeds the potential at which weathering can supply new soil.

An environmental factor regarded as being of prime importance in terms of landscape and slope development is climate. The fundamental concept of climatic geomorphology is that a unique assemblage of landforms are related to climate and to the interrelated factor, vegetation. If this is true, the control should be reflected in the slopes of an area, which should progress to different equilibrium forms through prolonged operation of different stable climatic conditions. There are two extremes of view regarding climatic geomorphology. Some geomorphologists would maintain that climate is of overriding importance and that all landscapes can be distinguished on climatic parameters. Conversely, others insist that climate has only a modifying influence and that landscapes tend towards common states of development, irrespective of climatic conditions.

It cannot be stressed too strongly that both of the extremes of viewpoint outlined above are unrealistic (and hopefully minority views). On the one hand, climate is only a single factor among many environmental influences which 'set the stage' for weathering and erosion on slopes. One might just as easily argue that geology is of paramount importance in determining what landscapes look like. On the other hand it would be foolish to argue that climate has no control over landscape in general and hillslopes in particular. Such a great range of conditions exists across the climatic zones of the world that it is unlikely that no differences in landscape development will be seen.

6.4.2 Important aspects of climate

So far we have only talked rather loosely about climate, and it is now necessary to define it a little more carefully.

The two most widely used parameters to describe climate are mean annual rainfall and mean annual temperature. Another important parameter can be derived from temperature and rainfall, namely evapotranspiration. Evapotranspiration is the amount of water lost from the earth's surface due to evaporation and plant respiration, and the difference between it and rainfall is the amount of water flowing on hillslopes, available for use in weathering and transport.

These parameters provide simple ways of characterising climates and for some places, perhaps the Congo Basin or the middle of the Sahara Desert, they may give quite a good impression of what conditions are like from day to day. In these areas, daily variations and seasonal variations in weather are *comparatively* small (note the emphasis on 'comparatively'; even in the places mentioned significant variations from mean conditions do occur). But in the British Isles, and a great part of the world's·land surface, mean annual conditions tell us practically nothing about what the weather is really like. Thus, mean temperature and rainfall are too coarse to be of much value in climatic description in many environments.

Climatic involvement in hillslope processes can be quite subtle. In addition to temperature and rainfall, issues such as number of frost cycles, or wetting and drying oscillations in the soil, rainfall intensity, and so on, may exert their influences in specific cases. However, it becomes very difficult to generalise the effects of climate at the detailed level and so most of the ensuing discussion is on the basis of temperature and rainfall. The inadequacies of such an approach should be realised.

6.4.3 Climate and dominant hillslope processes

Since we have taken a process approach throughout this book it is only logical to discuss the role of climate in the context of processes. Thus, for climatic differentiation of slope form to be a realistic proposition as far as we are concerned, we must be able to show consistently different balances of processes between the broad climatic types. Some climatic influences which are thought to have wide significance are discussed in this section, but an attempt to describe typical slopes from the whole range of climatic regimes is not possible in this short text.

An important primary control exerted by climate is its influence upon the balance between weathering and erosion. This balance determines the accumulation of regolith or soil on slopes. Other factors being equal (that is, in situations of similar geology, rates of stream downcutting, relief, and so on), more humid climates favour the accumulation of weathered regolith. The presence of water allows plant colonisation to proceed and stimulates the chemical weathering of bedrock. In arid regions chemical weathering and plant growth, which are strongly interrelated, are inhibited by lack of water, and so soils develop more slowly. Temperature is a further stimulus to the soil and vegetation system, with warmer regions favouring plant growth and chemical weathering. Regoliths accumulating over bedrock in colder and drier areas tend to be less chemically decomposed than their counterparts in warmer and wetter regions. The character of the regolith often has a direct influence on processes acting on slopes. For example it is generally true that the strength of a regolith decreases as its degree of weathering increases, which has an important part to play in the susceptibility of soils to landsliding.

A second area of climatic control, and probably the most significant in hillslope process terms, is the strong relationship between hydrology and climate. This has already been discussed extensively in Chapters 5.1 and 5.2, but requires reiteration here. Wet

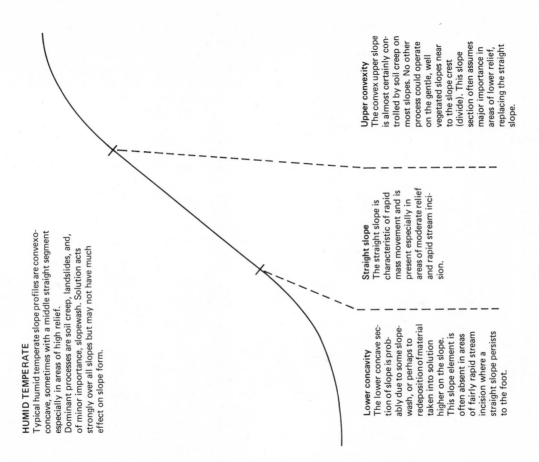

HUMID TEMPERATE
Typical humid temperate slope profiles are convexo-concave, sometimes with a middle straight segment especially in areas of high relief. Dominant processes are soil creep, landslides, and, of minor importance, slopewash. Solution acts strongly over all slopes but may not have much effect on slope form.

Upper convexity
The convex upper slope is almost certainly controlled by soil creep on most slopes. No other process could operate on the gentle, well vegetated slopes near to the slope crest (divide). This slope section often assumes major importance in areas of lower relief, replacing the straight slope.

Straight slope
The straight slope is characteristic of rapid mass movement and is present especially in areas of moderate relief and rapid stream incision.

Lower concavity
The lower concave section of slope is probably due to some slope-wash, or perhaps to redeposition of material taken into solution higher on the slope. This slope element is often absent in areas of fairly rapid stream incision where a straight slope persists to the foot.

Figure 6.22 A typical profile for humid temperate climates

183

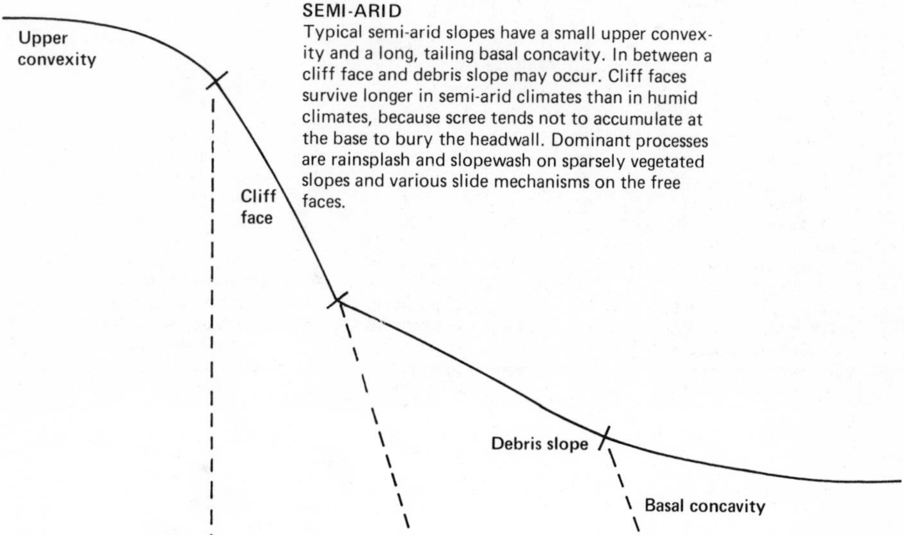

SEMI-ARID

Typical semi-arid slopes have a small upper convexity and a long, tailing basal concavity. In between a cliff face and debris slope may occur. Cliff faces survive longer in semi-arid climates than in humid climates, because scree tends not to accumulate at the base to bury the headwall. Dominant processes are rainsplash and slopewash on sparsely vegetated slopes and various slide mechanisms on the free faces.

Upper convexity

Cliff face

Debris slope

Basal concavity

Upper convexity
The upper convex slope in semi-arid areas is a rain-splash-controlled segment. On these poorly vegetated slopes rainsplash can only be dominant where little slopewash occurs. This is effectively confined to the area close to the divide, where overland flow depths are small.

Cliff face
The cliff face retreats by moderate scale rockwall processes such as wedge, slab or toppling failures. It supplies material to the debris slope below but is not usually buried by the accumulation. (See debris slope)

Debris slope
The debris slope of semi-arid areas is only superficially similar to scree slopes seen in humid temperate areas. Material supplied to the slope usually disintegrates quite rapidly under fairly intense mechanical weathering and is carried away by slopewash. Thus, only a thin veneer of boulders is present on the debris slope at any time, and so it does not accumulate to mantle the cliff face.

Basal concavity
A sharp break in slope exists between the basal slope and the debris slope. This marks the abrupt transition from rockfall debris to sediment transported and deposited by slopewash. Thus, the basal slope is formed by wash processes and may be the most significant part of many semi-arid slope systems.

Figure 6.23 A typical profile of a semi-arid slope

climates favour the development of thick soils and a good plant cover. This favours infiltration of water into the soil because plants promote an open, permeable soil structure, and because they reduce the velocity of surface flow, allowing a longer time for infiltration. Arid and semi-arid land soils have a sparse plant cover and tend to be much less permeable. Thus, surface overland flow is the dominant hydrological process on these slopes.

In terms of hillslope processes, humid regions are most prone to processes which are enhanced by the presence of soil water. On steeper slopes this includes landslides, where porewater pressures are very important (Chapter 4.2.1); and on gentler slopes one may expect soil creep to occur due to soil moisture fluctuations (Chapter 4.4). Chemical removal is of course ubiquitous on humid climate slopes. In semi-arid climates, processes which involve surface water flow are most dominant. *Figure 5.5* demonstrates how surface water erosion (slopewash) reaches its optimum in semi-arid regions. Rate of transport by surface water on bare slopes increases with rainfall. But plant cover also increases with rainfall, progressively protecting the slope from erosion until only very low rates occur on fully vegetated slopes. The maximum slopewash erosion therefore occurs in semi-arid climates where a combination of infrequent but intense rains and sparse vegetation makes the slopes most susceptible. Rainsplash is also important in semi-arid regions for similar reasons, though because it operates at slower rates than surface water flow erosion, it is only important close to the divides where there is little flow. Typical slope profiles for humid temperature and semi-arid climatic zones are shown in *Figures 6.22* and *6.23* with explanations of the dominant processes in different parts of the profiles.

It would be possible to discuss dominant hillslope form across all climatic zones. Apart from limitations of space, this exercise would be of dubious value, however. A great deal of process overlap occurs between climatic zones because, as already mentioned, climate is only one of the environmental factors influencing slopes.

CHAPTER 7   HILLSLOPES AND MAN

7.1 MAN-MODIFIED HILLSLOPES

7.1.1 Man — the most significant slope-modifying agency?

So far the main emphasis of this book has been on natural slopes, the processes which act upon them, and the forms they take. But in Chapters 1.2 and 1.3 it was mentioned that man has many interests in slopes, and influences them in many ways. Man has been an important modifier of the earth's surface for very many years and will continue to increase in importance as the scale of human enterprise increases. It is hardly surprising then, that man-made or modified slopes form a major part of some landscapes, and certainly form the bulk of the land surface of the developed world. Man's influence upon hillslopes could easily form the subject of a book in its own right, and there is only space here to talk very briefly about some of the more important changes which are made.

Man's influence on slopes extends from the building of completely artificial slopes to the slight, and often unintentional, modification of existing slopes. An obvious example of the former types of slope are the cuttings and embankments made in construction projects, such as roads and railways.

Cuttings are artificial features which expose undisturbed rock and soil to weathering and transport processes, and they probably come as close as is possible to a primary exposed surface. They are usually steep-sided in comparison to adjacent natural slopes in similar materials, and are therefore frequently markedly discordant landscape features.

Embankments are positive relief features which are constructed from newly placed material. Hence their internal structure depends on how they are constructed, and not upon natural geological or geomorphological processes. The visual impact of embankments is no less striking than cuttings. Their steep-sided, sweeping forms often contrast sharply with the natural landscape.

Other major categories of artificial slopes include quarry faces and spoil heaps, which are the product of mining industries. Finally, many coastal slopes require protection against coastal erosion, and substantial remodelling of the cliff profile is necessary. Consequently many protected coastal slopes can be regarded as artificial slopes.

The examples in the preceding paragraph are very obvious examples of landscape changes by man. But much less obvious and yet more widespread changes may occur because of agricultural practice. Some readers may find it surprising that there is scarcely any land in the British Isles which can be regarded as in its natural state, but practically all has undergone some form of agricultural pursuit at some time. The degree of modification varies from total vegetation change to minor modifications occasioned by casual grazing. Another, more serious consequence of agriculture is the problem of accelerated erosion which has already been referred to in 1.3.2 and is discussed in more detail in Chapter 7.2. With few exceptions this has rarely been a problem in Britain, though over much of the world very serious erosional modification of hillslopes has resulted from poor agricultural practice. Even in the pursuit of leisure, man causes important changes in the landscape. Increased tourist pressure on footpaths, and on sensitive environments such as coastal dunes, has resulted in considerable deterioration of the vegetation cover, and subsequent erosion. Many of these changes are entirely unintentional by-products of human activity, yet they have far-reaching effects in the natural landscape.

The rest of this section is concerned with aspects of artificially made slopes, and especially with aspects of their successful function. Chapter 7.2 concentrates upon the importance of land use in soil erosion, and Chapter 7.3 examines the possibilities for learning lessons from natural hillslopes which will help us to develop and manage slopes more successfully.

7.1.2 Problems of cutting and embankment slopes

All artificially constructed slopes change the regime of processes acting at that point on the earth's surface. Consequently, one should expect that these slopes are potentially susceptible to processes which do not operate, or which operate much more slowly, on adjacent natural slopes. Accordingly, the design of artificial slopes should be sympathetic to this concept and should seek to lessen the risk of unfavourable hillslope processes occurring.

Cuttings and embankments are usually constructed to steep angles, at least in comparison to nearby natural slopes on similar materials. The most serious threat to their

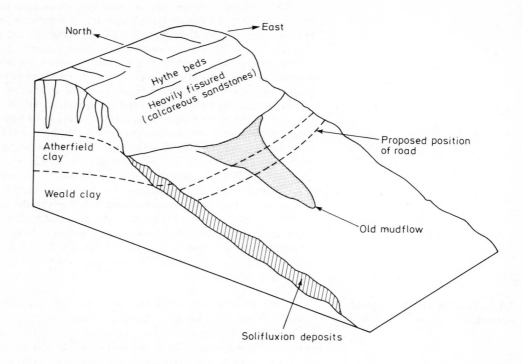

North → ← East

Hythe beds
Heavily fissured
(calcareous sandstones)

Atherfield
clay

Weald clay

Proposed position
of road

Old mudflow

Solifluxion deposits

**Figure 7.1 Geology and topography of
the Hythe Beds escarpment near
Sevenoaks (After Symons and Booth, 1971)**

function is therefore the risk of landsliding, due to instability in the soil or rock from which they are formed. Pre-existing shear planes from previous landsliding events present the most limiting condition. If old landslides are crossed by cuttings, embankments, or any other structure, reactivation of the old slide is a real possibility. Strength of the soil or rock is always lowest along a pre-existing shear plane, and you may recall from Chapter 4.2 that this lower strength is called the residual strength.

A good example of this problem occurred during the construction of the Sevenoaks bypass in Kent. The road was constrained to climb the escarpment of the Hythe Beds (Lower Greensand), which in this area are sandstones and sandy limestones. These harder beds are underlain by the softer strata of the Atherfield and Weald Clays (*Figure 7.1*). It was not realised when the road was planned that the route passed over a number of inactive low-angled mudslides. These slides, which occurred in the Pleistocene when the ground was frozen, form low-angled, hummocky lobes at the foot of the scarp. Problems were encountered in an embankment, which was displaced by reactivation of a mudslide, and in a cutting where another lobe began to advance onto the carriageway. *Figure 7.2*

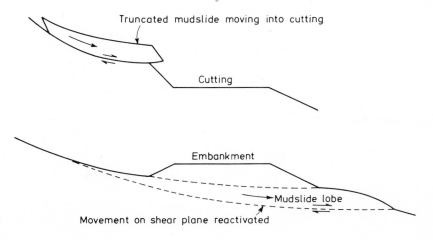

Truncated mudslide moving into cutting

Cutting

Embankment

Mudslide lobe

Movement on shear plane reactivated

Figure 7.2 Schematic diagram to show types of failure on the Sevenoaks bypass

shows schematically the nature of the failures which affected the road during construction. A realignment of the road was necessary to avoid the majority of the mudslides and redesign of the embankments and cuttings where a crossing could not be avoided (Symons and Booth, 1971).

Even if no previous shearplanes are present in the soil, a careful study is necessary before cuttings or embankments are built. Sometimes it is possible to design a cutting with respect to the intact strength (Chapter 4.2.1) of the soil. This means that the whole strength of the soil as measured can be relied upon to operate on a fairly long-term basis, and the steepest possible angle for the cutting sides may be used. Frequently however, there are reasons why the full strength cannot be relied upon. The problem of shearplanes has already been discussed above. Other important factors which reduce the strength of soils and rocks are weathering, fissures and joints, and high porewater pressures. Very detailed information and a careful survey of materials, topography, and hydrology must therefore be available to the engineer, in order that a satisfactory design may be accomplished.

Another important problem associated with artificially cut or built slopes is that of surface water erosion. During and immediately after construction the slopes are entirely free of vegetation and, as was explained in detail in 5.2.1, vegetation is paramount in restricting erosion by flowing water. Accordingly, steep, unvegetated artificial slopes are seriously at risk with respect to gully erosion, especially if they are constructed from coarse silt- or fine sand-size material. Because of this, every effort should be made to establish a good vegetation cover on newly constructed slopes as soon as possible.

7.1.3 Cuttings in hard rocks

Cuttings through hard rock may well be the most troublesome and unpredictable of all artificial slopes. Recall from 4.2.2 that the most common failures in hard rocks are controlled by the presence of joints and other discontinuities. To minimise the risk of slides by wedge, slab or toppling mechanisms it is essential that the angle and orientation of the cutting faces do not unfavourably intersect the discontinuity pattern. Overall geological structure and the disposition of different lithologies can also be important and there have been many examples of failures on hard rock cuttings which have been related to the broad geological picture.

Artificial cutting

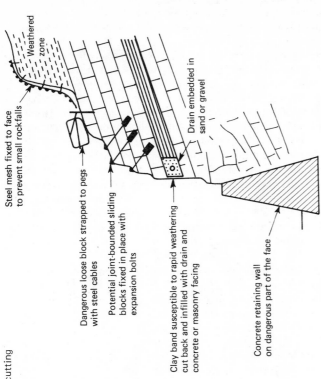

Steel mesh fixed to face
to prevent small rockfalls

Weathered zone

Dangerous loose block strapped to pegs
with steel cables

Potential joint-bounded sliding
blocks fixed in place with
expansion bolts

Drain embedded in
sand or gravel

Clay band susceptible to rapid weathering
cut back and infilled with drain and
concrete or masonry facing

Concrete retaining wall
on dangerous part of the face

Natural slope

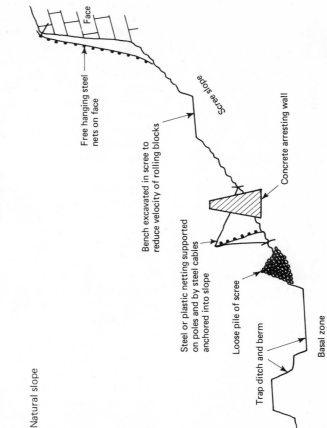

Face

Free hanging steel
nets on face

Scree slope

Bench excavated in scree to
reduce velocity of rolling blocks

Concrete arresting wall

Steel or plastic netting supported
on poles and by steel cables
anchored into slope

Loose pile of scree

Trap ditch and berm

Basal zone

Figure 7.3 Some measures to reduce the rockfall hazard in cuttings and on natural rock slopes

A good example of such a slide was encountered in a cutting on the Cullompton bypass (Devon), which is part of the M5 motorway. This cutting, which is almost 50 m deep, traverses upper Carboniferous rocks. A large, deep-seated slide took place immediately after construction, necessitating regrading of the cutting side. The slide was very complex and it is not possible to describe in detail the many factors which played a part in its initiation. However, flow of groundwater through permeable sandstone beds, confined by adjacent mudstones, was of key importance. The problem of groundwater flow was heightened by the synclinal structure of the rocks, since water flow was concentrated into the centre of the syncline and directed into the cutting. This landslide clearly demonstrated how important it is to fully understand the geological structure, lithologies, and groundwater hydrology of a site before the design stage.

Unlike cuttings through soft rocks and unconsolidated deposits, many hard rock cuttings are constructed to steep angles and left with considerable bare rock exposure. The rock slope is therefore exposed to weathering, introducing the possibility of degradation and rockfall of material from the slope. Hence it becomes important to monitor the behaviour of rock cuttings after construction, and to introduce measures to control rockfall if necessary. Many techniques can be used to restrict rockfall and to catch fallen blocks and the main ones are illustrated in *Figure 7.3*.

7.1.4 Embankments and embankment dams

Embankment construction is far less susceptible to variability of the natural material, provided of course that the foundation on which the embankment is built is free from problems. The embankment itself is built from relatively homogeneous quarried rock or excavated soil, whose behaviour can easily be studied experimentally. Having said that, a casual glance at motorways in Britain will be enough to assure most people that problems of stability are by no means infrequent on embankments. However, one special type of embankment, the embankment dam, is worthy of extra comment because it must resist water pressure forces owing to the contained reservoir.

Embankment dams are built from hard rock debris (rockfill) or softer soil debris (earthfill), and they are particularly suitable in poor foundation areas where settlement of the dam is likely. Such settlements might spell disaster for a concrete dam, but the flexibility of an embankment allows this problem to be accommodated. But since

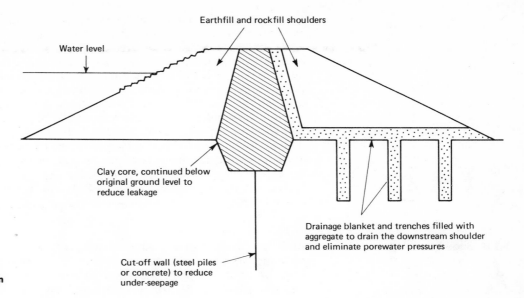

Earthfill and rockfill shoulders

Water level

Clay core, continued below
original ground level to
reduce leakage

Cut-off wall (steel piles
or concrete) to reduce
under-seepage

Drainage blanket and trenches filled with
aggregate to drain the downstream shoulder
and eliminate porewater pressures

Figure 7.4 A typical embankment dam

embankment dams are made of unconsolidated, permeable materials, they suffer seepage through their structure. Dam design must minimise seepage to reduce water loss by leakage, and to eliminate high porewater pressures in the downstream face of the dam, which would quickly lead to collapse. A typical embankment dam is illustrated in *Figure 7.4*. Leakage is reduced by the relatively impermeable clay core, while porewater pressures in the downstream shoulder are controlled by coarse drainage blankets within and beneath the structure.

7.1.5 Quarry faces — artificial slopes of the mining industries

Quarries and spoil heaps are a major category of artificial slopes in our landscape. Quarry faces are steep rock or sand and gravel slopes, which are totally unvegetated and unadjusted to the natural geomorphic processes operating in the landscape. Spoil heaps too, are steep-sided features and often markedly discordant to the natural slopes. In common with quarries, spoil heaps have a more haphazard and unplanned appearance than most other artificial slopes. However, this undesigned appearance is often illusory. While it is fair to say that many old spoil heaps and quarry faces were developed casually, modern examples are carefully designed to ensure the continued successful operation of the workings.

The main problem associated with quarry faces is again one of stability. Although long-term stability may be unimportant to the quarry operator, short-term stability can pose serious problems. Especially important are the continued viability of the quarrying operations and the safety of faces that personnel must work beneath. Serious failures of faces or benches in a quarry can disrupt production, possibly by removing haulage-ways or damaging machinery. Once again, careful planning may be necessary to minimise the risk of failures along discontinuities which are inclined towards the face.

7.1.6 Spoil heaps

Wherever extraction of a mineral resource takes place, tipping of spoil is unfortunately necessary. Most spoil in Britain results from coal mining, but power station fly ash, kaolinite waste, slate and stone quarrying, and metalliferous mining have also contributed large quantities. Typically, spoil heaps have been built at the 'lower angle of repose' (Chapters 4.2.1 and 4.2.6) of the waste debris. That is, they have been tipped from a tramway or aerial ropeway, and the sideslopes allowed to accumulate at the angle at which the tipped debris comes to rest. Consequently, very steep slopes are constructed, with little regard to the potential hazards which may develop. Actually, high, steep-sided tips are not nearly so often used today, and most problems are encountered with older ones which were constructed when some aspects of tip stability were not understood.

Often the most obvious problem associated with tips is gully erosion. Their steep, unvegetated sides render them highly prone to surface water erosion, resulting in deep gullies and unsightly waste deposition over surrounding land and in streams. Vegetation, the key to controlling water erosion on slopes (Chapter 5.2.1) is often difficult to establish, due to slope steepness and low fertility in many tip materials. Coal tips, for example, tend to become acid through time, and metalliferous mine waste is often highly toxic to plants.

Many tips display serious stability problems, either in the foundations owing to the load of tipped material, or in the tip itself. While small-scale failures of the tip may not be significant, larger failures may threaten adjacent structures and property and be of some concern. But the most serious problem of tip stability is their susceptibility to flowslide behaviour. This problem was highlighted by the tragic Aberfan coal tip flowslide (South Wales) in 1966, in which 116 schoolchildren and 28 adults lost their lives. Hitherto flowslides on tips had received little attention, even though their occurrence was known over 50 years previously.

Flowslides (Chapter 4.3) occur when slipped debris becomes highly mobile. As discussed in Chapter 4.3, the most usual immediate cause is a high water content and high porewater pressure within the debris. Many tip materials are prone to flowslide behaviour because they possess large percentages of silt and fine sand debris, in addition to coarser particles. Also many tipped materials, including coal tip waste, are prone to weathering and breakdown through time and a consequent reduction in strength. Finally, tipped debris is usually very loosely placed and is likely to reduce in volume when disturbed by landsliding. When the material is saturated, this reduction in volume can tend to expel porewater and lead to high porewater pressures.

The Aberfan flowslide illustrates clearly the process of flowsliding, and the importance of an initial trigger mechanism. The geology of the Aberfan tip site is shown in *Figure 7.5*.

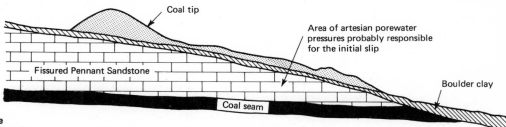

Figure 7.5 Schematic diagram of the Aberfan coal tip site (After Bishop, 1973)

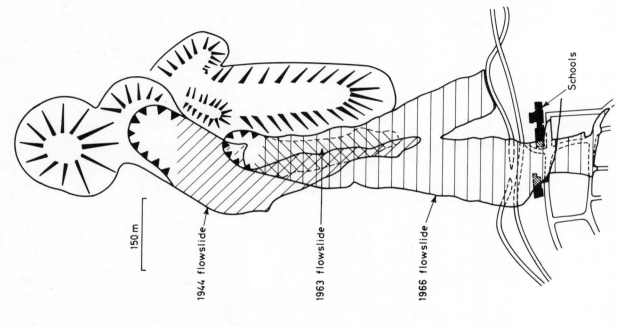

150 m

1944 flowslide

1963 flowslide

1966 flowslide

Schools

Figure 7.6 Sketch map to show the extent of the 1966 and earlier flowslides at Aberfan (After Bishop, 1973)

196

It can be seen that the highly fissured, permeable Pennant Sandstone forms a confined aquifer between a coal seam below and a surface covering of boulder clay, and caused artesian water pressures to build up beneath the toe of the tip. These water pressures probably caused a small slip in the boulder clay at the base of the tip, which in turn triggered the flowslide. After the initial failure, the debris mobilised and flowed downslope for about 600 metres on a slope of 12.5°. A considerable volume of water flowed from the sandstone aquifer, released by the stripping of the boulder clay cover, though it followed the main flowslide and played no part in its operation. The full extent of the flowslide is shown in *Figure 7.6*. It is difficult to decide what was responsible for the behaviour of the Aberfan tip, but several factors seem to have been important. First, the initial density of the tipped material was very low and so it was susceptible to the collapsing behaviour and high porewater pressures mentioned above. Secondly, the material tended to break down with shearing, resulting in a progressive lowering of its frictional strength as it moved downslope.

After the Aberfan flowslide, many coal tips in South Wales were examined and a large number were regraded to increase their safety. Also, the practice of building steep, conical tips on sloping ground has largely been replaced by extensive, flat tips.

7.1.7 Coastal protection sites

Coastal cliffs in soft rocks, such as the London Clay of the Thames estuary, may be receding at very rapid rates because of basal marine erosion. London Clay cliffs on the Isle of Sheppey (North Kent) for example, are currently receding as fast as 1 m yr^{-1}. Nearby, parts of the town of Herne Bay have been threatened by similar rates of recession for many years. Protection works were started in 1938 and extended in 1968, and consisted of regrading the landslide zone to a stable angle, incorporating drains into the cliff and foreshore area to reduce porewater pressures, and protecting the cliff foot with a sea wall and promenade. *Figure 7.7* hows a simplified sketch of the Herne Bay protection measures alongside a typical profile of the cliff prior to stabilisation. The Herne Bay scheme illustrates quite well the important issues of coastal protection; namely, basal protection, regrading, and drainage. Many similar schemes exist on other cliffs composed of soft rocks in the British Isles and throughout the world.

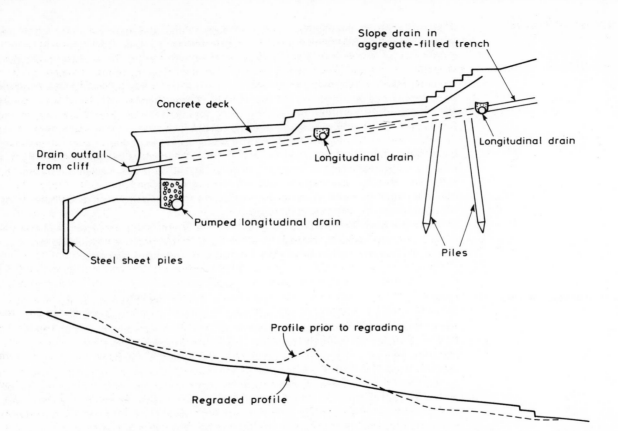

Figure 7.7 Measures taken to control cliff recession on London Clay Cliffs (After Allan, 1968; and Bayley, 1972)

7.2 LAND USE AND SOIL EROSION

7.2.1 The soil erosion problem

The basic premise of this book, which we have restated frequently, is that under natural conditions soil formation and erosion are in a state of balance, in which the natural vegetation cover is an essential factor. Any interference inevitably leads to accelerated erosion (Chapter 1.3) and there is no way it can be prevented. All that we can do is regulate land use and land management practices so as to try to keep accelerated erosion to a minimum. The history of agriculture is littered with horror stories of soil erosion from the very earliest times right up to the present day. Perhaps the best-known case is the 'dust bowl' in the central plans of the USA in the 1930s. Like many other cases of serious erosion this was not just a physical disaster but also a great human tragedy as has been so movingly described by John Steinbeck in his novel *The Grapes of Wrath*.

It would be satisfying to be able to say that soil erosion is no longer a serious problem but unfortunately this is not the case. From the purely scientific point of view, the nature and extent of the problem and the measures necessary for dealing with it have been well known for a long time, although research continues and methods are constantly being improved. The real problems at present are financial and sociological. It is often very difficult to persuade farmers to take the steps necessary to reduce soil erosion.

At its worst, erosion leads to the complete loss of the soil, but more commonly just the topsoil is lost. Removal of the topsoil reduces the infiltration capacity of the soil so the amount of soil wash is increased and runoff is greater. Part of the water which is lost as runoff would otherwise be stored in the soil and be available to plants. The micro-fauna and flora of the soil are concentrated in the topsoil where they decompose organic litter and release nutrients for recycling in the vegetation. Stripping of the topsoil there-fore greatly reduces nutrient recycling. Good agricultural land on lower topographic sites can be buried by poorer material eroded from upslope and so the better land, even though not eroded, is still damaged.

Soil erosion also adds to pollution. Where soil erosion is occurring, streams will have an increased load of particulate matter, which must be removed before the water can be used for human consumption. This is an expensive operation. Soil eroded from slopes may be deposited in reservoirs, reducing their storage capacity and in the most extreme cases (of which there are many) entirely filling the reservoir and turning it into a water-fall. Increased runoff from eroded land takes with it fertilisers and pesticides applied by farmers. This causes problems for farmers since they are not getting the full benefit from these expensive products and it also adds to the chemical pollution of water supplies.

Nitrate pollution produced in this way is a direct health hazard since it can cause a condition in infants known as methaemoglobinaemia in which the ability of the red blood cells to carry oxygen is impaired leading to dizziness, headaches, diarrhoea and anaemia. High nutrient levels (for example nitrates and phosphates) in runoff can also lead to eutrophication of lakes and rivers. The nitrates and phosphates are plant nutrients which stimulate growth of aquatic plants, especially algae. These plants and their decomposing remains deplete the dissolved oxygen supply in the water which leads to a deterioration of the aesthetic and life-supporting qualities of the water.

It is possible to compare rates of erosion and runoff for different land uses but for the comparison to be meaningful all factors other than land use must be held constant. This may be done in two ways; by measuring runoff and sediment load from controlled experimental areas; or by directly measuring ground loss by erosion or historical deposition from a known area. Numerous controlled experiments have been carried out and although the precise details and rates vary from one area to another, the basic principles are inevitably the same. An example is shown in *Figure 7.8*. The soil in this case was silt loam on a 7° slope. Under undisturbed woodland, surface runoff and soil loss were negligible. Runoff from grassland was greater but soil loss was still very small. Wherever cultivation is practised, runoff and soil loss increase dramatically as shown for rotation, bare fallow and cotton in *Figure 7.8*. Compare these results with the infiltration measurements under forest and pasture shown in *Figure 5.3*. An example of deposition from accelerated erosion is shown in *Figure 7.9*. The buried fence in that photograph is 35 years old.

The problem of accelerated soil erosion is not confined to agricultural land, though clearly this represents the major part of the problem because of the area covered. Erosion related to construction in urban and suburban areas has been widely documented but it is usually only temporary and ceases when the construction is complete. Severe erosion may also occur during road construction and if water draining from the road surface is not adequately controlled, prolonged and serious gullying can occur along the margins of the road. When forest timber is being harvested it is necessary to construct temporary roads (snig tracks) to get the timber out and these frequently become badly eroded.

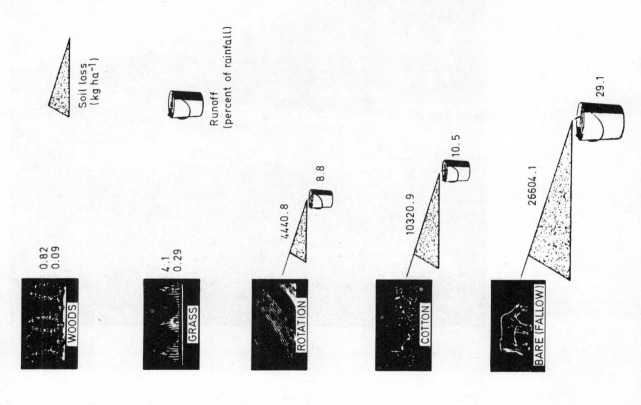

Soil loss
(kg ha^{-1})

Runoff
(percent of rainfall)

WOODS
0.82
0.09

GRASS
4.1
0.29

ROTATION
4440.8
8.8

COTTON
10320.9
10.5

BARE (FALLOW)
26604.1
29.1

Figure 7.8 Annual losses of soil and water from the same kind of land with the same rainfall, under different land uses (After Bennett, 1939)

Figure 7.9 Wooden fence posts buried by sediment produced by accelerated erosion.

Footpaths in National Parks and other areas frequently visited by the public can also become sites of rapid erosion. In all these cases techniques are available for minimising erosion; in all too few are they used.

7.2.2 Soil conservation

Soil conservation is the wise use of soil so as to ensure production indefinitely at the maximum level for which the land is suitable. The most disastrous cases of soil erosion occur where land is cultivated in areas totally unsuited for it. This was the case in the 'dust bowl' mentioned above where the only reasonable conservation practice was to cease arable farming and return the land to grass. In less extreme cases erosion can be kept within acceptable limits by the use of appropriate soil conservation practices. These fall into two broad groups: agronomic and mechanical.

Agronomic measures are those involving land management and cropping practices. The most important of these is to ensure that the land is used in accordance with its capability. The choice of crops and cultivation techniques should be aimed at maintaining soil structure and fertility. A well structured soil promotes infiltration, thus reducing runoff and the risk of erosion. If the soil surface is left bare during heavy rain, rainbeat can break up the structures and seal the soil surface (Chapter 5.2), thus leading to increased runoff and erosion. Cover crops can be grown in the period between harvesting and planting cash crops and then ploughed in as 'green manure' to become part of the organic matter of the topsoil. Leaving crop stubble on the field, or partly ploughing it in (stubble mulching) also protects the soil surface.

It is a common practice for farmers to burn crop residues to facilitate seed bed preparation. Burning is also popular among foresters as a means of preparing a harvested area for replanting. In nearly all cases, burning leads to increased erosion and to loss of nutrients and these problems are well known. Despite this the practice persists because it is quick and convenient. Overgrazing pastures reduces the vegetation cover and leaves part of the soil surface bare leading to accelerated erosion. Reduced stocking rates or planting species which grow more vigorously (improved pastures) are obvious means of alleviating the problem.

Mechanical measures for combatting soil erosion are those requiring special structures and field designs. There are a large range of these to suit local conditions and specific problems. Cultivation along the contour is used to aid local retention on the soil surface of rainfall which temporarily exceeds infiltration capacity. Cultivated strips are sometimes separated by strips retained in permanent pasture to promote infiltration. On steeper land and heavy soils, cultivation along the contour may not be enough to prevent surface runoff so contour banks are constructed to trap runoff and divert it into permanently grassed waterways (*see Figure 7.12*). Conservation projects often include small dams to retain excess runoff for watering stock. Great care is needed in the design of mechanical conservation structures since the failure of contour banks may lead to worse erosion than would have occurred had they not been built. The effect of contour ploughing and contour banks is to reduce effective slope length. Where cultivation terraces are built, slope angle is also reduced. As we have seen from the Universal Soil-Loss Equation (Chapter 1.3), slope length and angle are important factors in erosion.

Mechanical and agronomic measures are usually complementary and both are necessary in combatting erosion. The real problem lies not so much in knowing how to design and implement conservation programmes, but in getting farmers, foresters and others concerned in land management to realise that it is in their own best interests to implement them. The Universal Soil-Loss Equation (introduced in Chapter 1.3) is best used as a means of determining appropriate cropping and conservation practices to keep soil loss to a minimum. An example of the use of the equation is given below.

7.2.3 The Universal Soil-Loss Equation: An example from Iowa (Hayes, 1977)*

The example we will use is a 160 acre (65 hectare) farm in Iowa, USA, the location of which is shown in *Figure 7.10*. The soil on the farm is a moderately eroded silt loam on 6% (3.5°) slopes. The effective slope length is 300 feet (92 m) under the present cultivation practice of ploughing up and down the slope. Maize is grown continuously with conventional tillage practices, ploughing in the autumn and leaving the land in bare fallow prior to planting in the spring. The crop residue (stalks and leaves) is ploughed under. The farmer wishes to reduce soil loss from his land and consults a soil conservation officer who uses the Universal Soil-Loss Equation to estimate the rate of soil loss under the present management system and under proposed alternative systems.

The Universal Soil-Loss Equation was presented in Chapter 1.3. Clearly the equation can only be used in locations where numerical values of the factors are known or can be determined. The equation was developed in the USA from the results of numerous controlled experiments on field plots and small watersheds over a period of more than 20 years. This research has at least provided approximate values of the factors in the equation for the major agricultural regions of the USA and these are published as tables

---

*In this example the units acres and feet, with slope given as a percentage, will be used since all the original tables and nomograms are in these units.

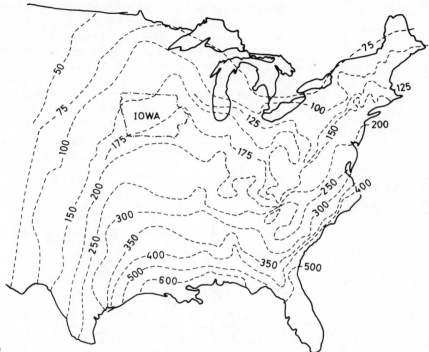

Figure 7.10 Values of *R*, the rainfall erosion index for the eastern United States (After Wischmeier and Smith, 1965)

and charts. The principles of the method are universally applicable and numerical values for the factors have recently been produced for several African countries.

The first two factors are fixed for a particular soil in any given location. The rainfall erosion index (R) is derived from the total kinetic energy of a storm multiplied by its maximum 30 minute intensity. These values are calculated for each storm over a period of as many years as possible and the long term annual average is R. Values of R for the

205

eastern United States are shown in **Figure 7.10**. It can be seen from that illustration that our example has an **R** value of 175. The soil erodibility factor (**K**) is also independent of other terms in the equation and for the silt loam soil in the example is 0.37. This is an empirical value derived from research station experiments.

Slope length (**L**) and gradient (**S**) are combined in the equation as the topographic factor, **LS**. It is interesting to note that this topographic factor is for a uniform (straight) slope and the effects of convexity or concavity have yet to be evaluated. This term is not independent since it will vary with erosion control practices. The installation of contour banks, for example, has the effect of reducing the effective slope length to the

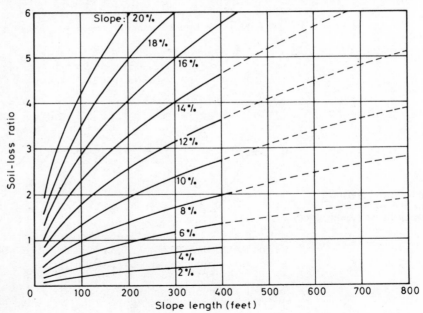

Figure 7.11 Chart for determining the topographic factor, LS (After Wischmeier and Smith, 1965)

distance between the contour banks. The factor *LS* for our example is read from the chart in *Figure 7.11* and for the present practices is 1.2. It was considered reasonable to install contour banks 120 feet (37 m) apart, thereby reducing effective slope length from 300 feet to 120 feet. This reduces *LS* to 0.74. (Note that the gradient remains unchanged).

The next step is to decide on 'soil loss tolerance' which is the minimum rate of soil erosion that will permit a high level of crop productivity to be sustained economically and indefinitely (*see* definition of soil conservation in 7.2.2). For soils in the USA, soil loss tolerance is considered to be in the range 1–5 tons acre^{-1} year^{-1} and for the soil in this example, the maximum permissible loss is 5 tons acre^{-1} year^{-1}. It is therefore necessary to find a combination of crop management factor (*C*) and erosion control practice factor (*P*) which will keep soil loss per unit area below 5 tons acre^{-1} year^{-1}. *P* and *C* are also taken from site-specific tables. The farmer then selects the land use and management combination best suited to his own needs.

TABLE 7.1 APPLICATION OF THE UNIVERSAL SOIL-LOSS EQUATION FOR PRESENT PRACTICE AND TWO ALTERNATIVE DESIGNS FOR A FARM IN IOWA, USA. (AFTER HAYES, 1977)

	R ×	K ×	LS ×	P ×	C	= A	
1	175	0.37	1.2	1.0	0.45 = 35		Present practice, up and down cultivation, autumn ploughing
2	175	0.37	1.2	1.0	0.35 = 2.4		Up and down cultivation, no autumn ploughing
3	175	0.37	0.74	0.5	0.03 = 0.7		Contour cultivation, contour banks, no autumn ploughing

In *Table 7.1* the equation has been solved for the farmers present practices and the annual average soil loss is 35 tons acre^{-1} year^{-1}, (line 1, *Table 7.1*). By simply deleting autumn ploughing (line 2, *Table 7.1*), the soil loss can be reduced to 2.4 tons acre^{-1} year^{-1}.

Figure 7.12 Vertical aerial photograph of the area around Pilton on the Darling Downs, south-east Queensland. Aerial photograph supplied by the Surveyor General, Queensland, and reproduced by arrangement with the Queensland Government. Crown copyright reserved

0   500   1000 metres

The reason for this drastic reduction in soil loss is that without autumn ploughing the crop residue remains on the soil surface throughout the winter, protecting it from rainsplash and inhibiting surface runoff. If in addition cultivation is carried out along the contour and contour banks are built the soil loss can be further reduced to 0.7 tons acre^{-1} year^{-1} (line 3, *Table 7.1*). Since both these alternatives are within the soil loss tolerance the farmer can choose the one he prefers.

The predictions of the equation are not for any particular year but are our indication of expected long term soil loss, where 'long term' is interpreted as being a period of the order of 20 years.

Given the dramatic reduction in erosion that can be brought about by relatively simple soil conservation measures, as shown by this example, it is remarkable that so many farmers continue to allow severe erosion to occur. *Figure 7.12* shows a small area of the Darling Downs in Southeastern Queensland where soil erosion is a serious problem. If you examine the photograph carefully, you will find a number of areas where severe gullying is occurring. Compare them with adjacent areas where contour banks have been built and contour cultivation is practised. The buried fence shown in *Figure 7.9* is located at the point arrowed on *Figure 7.12*. In this small area there are some farmers who use sound soil conservation practices and some who do not, and this emphasises the point made earlier that the reduction of erosion is as much a sociological problem as a scientific one.

7.3 LEARNING FROM NATURAL HILLSLOPES

Hillslopes are places of action. Processes of weathering and material transport operate on them to develop and modify the forms which we see in the landscape. Throughout this book it has been held that understanding of processes can be gained from basic mechanics and chemistry, and that the processes are the key to the development of the slopes. Natural hillslopes are usually well-adjusted to the environment in which they are developing. That is, their form, processes, vegetation, and so on, depend upon environmental factors such as climate, basal erosion, and geology. One could call the collection of slope attributes (form, process, vegetation etc.) the *appearance* of the slope. Slopes rapidly attain an equilibrium appearance, determined by the wider environment, and then show relatively little change. This is not to say that no further development or action is occurring, simply that the overall appearance of the slope system remains much the same.

We have seen in previous sections of this chapter that engineering or agricultural works potentially upset the natural balance between slope attributes. This is to be expected because all attributes are interrelated, and a change in one will inevitably lead to changes in the others. Frequently, the result is a change in processes operating on the slope, or an increase in the rate of some of those already operating. A new cutting can suffer from high rates of surface water erosion due to its steep, unvegetated slopes. The important interactions which are modified in this case are the removal of vegetation, changing the hydrology of the slope from predominantly subsurface to predominantly surface flow, and increasing the slope angle increases the erosive capacity of surface water flow. Cuttings too, may suffer from instabilities, due to the increase in slope angle. Agricultural practices also upset the natural balance through changes in the vegetation cover. Accelerated erosion (Chapters 1.3.2 and 7.2) is sometimes the result, with the loss of good agricultural land.

It seems that any human activity on slopes can result in an unfavourable reaction in the natural landscape. The aim of the engineer and agriculturalist should be to predict the reactions, and to minimise their effects. This may only be achieved by careful analysis of the mechanics of slope processes, and a sympathetic evaluation of the geology, hydrology, geomorphology and biogeography of the slopes. Potential problems may be built in or avoided, and we may therefore learn lessons from the natural slopes which will help us to manage artificial ones. We shall illustrate this principle with just two case histories; the design of an embankment and cutting across an old landslide zone, and a site investigation for a new road in mountainous country.

7.3.1 An embankment and cutting on an old landslide (Realignment of the A606 road in Leicestershire; Chandler, 1976, 1977)

When Rutland Water reservoir was impounded, part of the A606 road in Leicestershire was flooded and had to be diverted. The diversion required the road to climb a valley slope cut in Lias Clay, across the site of a large, presently inactive, landslide. *Figure 7.13* shows the extent of the landslide zone with respect to the road works. It can be seen that the road is upon slipped material for much of where it climbs the slope. A detailed

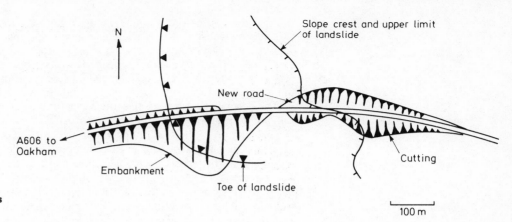

Figure 7.13 New section of A606 across landslide zone (After Chandler, 1977)

N

Slope crest and upper limit of landslide

New road

A606 to Oakham

Embankment

Toe of landslide

Cutting

100 m

investigation of the geomorphology and geology of the slope was carried out, using bore-holes and trial pits, and enabled the section shown diagrammatically in *Figure 7.14* to be constructed. The investigation showed landslide debris over much of the slope, thickening

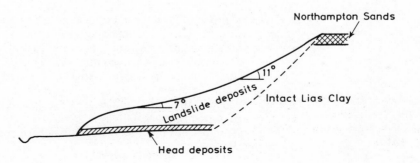

Figure 7.14 Diagrammatric sketch of landslide slope across which A606 road was routed

Northampton Sands

11°

7°

Landslide deposits

Intact Lias Clay

Head deposits

211

towards the base before thinning out again onto the footslope. Shearplanes were present beneath the landslide debris, showing conclusively that sliding had occurred in the recent geomorphological history of the slope. Significantly, the landslide had run out onto the footslope, and buried a layer of 'head' deposits. These consist of fragments of harder rock in a sandy/sandy-clay matrix, and represent a zone of higher permeability beneath the toe of the slide. This head layer became important in the final design of the crossing.

The normal procedure to find the strength of a soil would be to take samples of the soil, and to perform tests in the laboratory. The strength determined would then be used in a calculation of the stability of a proposed cutting or embankment. But because a landslide was already present in this example, Chandler reversed the procedure. He assumed that the landslide was *just* stable, and then calculated the strength of the clay material along the shearplane which would be necessary to maintain the slide mass in position. This was not an unreasonable assumption because the landslide had quite a fresh appearance. The value of strength, calculated entirely from field considerations and not laboratory measurements, was used in the final design of the road works.

An important part of the road crossing design was a series of drains, to lower the water table within the landslide. The investigation of the slope showed that the more permeable layer of head deposits was already acting as an inefficient drain, transmitting water from the base of the landslide debris. Further drains were installed to tap the water in the landslide and to link with the layer of head deposits, thus improving its efficiency as a drainage layer.

This road project shows the importance of a very good investigation of the slope prior to design work starting. But more importantly it shows how aspects of the natural slope were used in final design, integrating the works as well as possible with the natural slope regime. First, the landslide itself was used to calculate the strength of the clay. This meant that one could be fairly certain that the calculated strength could be relied upon to operate in the field, unlike laboratory determinations which often overestimate field-strength. Secondly, the natural drainage of the slope was integrated with the drainage incorporated as part of the stabilisation works. Had the initial investigation been less rigorous the presence of the head deposits may have been overlooked, to the detriment of the final design.

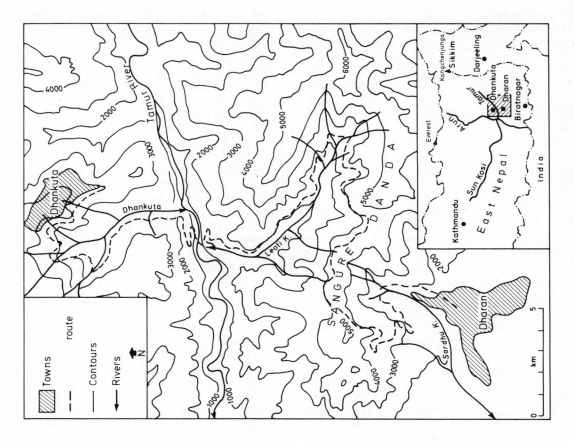

Figure 7.15 Location of the Dharan–Dhankuta road project (After Brunsden et al., 1975)

213

7.3.2 The site investigation for a road through Himalayan terrain in Nepal (Brunsden *et al*., 1975)

The projected road between Dharan and Dhankuta in Nepal (*Figure 7.15*) passes through extremely difficult Himalayan terrain, and illustrates well the principle that we must learn to have regard for the condition of natural slopes when planning any civil engineering project. The route makes three ascents and descents of over 1000 m each, is forced to cross long, steep slopes, and to travel through deeply incised gorges. Often the slopes are mantled with highly weathered soils or screes and, with the continued high rates of vertical erosion by rivers, slope instabilities in rock and soil are very frequent. The difficulties of finding a satisfactory route from the point of view of gradient alone in this sort of country are immense, a point well illustrated by the fact that a projected route length of 65 km is necessary to link these two towns, which are 18 km apart on the map.

Since this particular road project was regarded as low cost, and because very detailed site investigation would be difficult in this area, field mapping of geomorphic features and processes was chosen as the main site investigation technique. The mapping was quite detailed and included information on sediments, bedrock exposure, drainage, slope instabilities, breaks of slope and other geomorphic features. A simplified version of part of the finished map is shown in *Figure 7.16*. A very great deal of information has been omitted from the diagram and only details relating to slope instabilities and gullying have been included. It is readily appreciable from this small extract of the total route that severe problems are present, which could threaten the long-term viability of the road. The projected route is cut by many actively eroding gullies and traverses numerous active landslide features, and it is clear that substantial relocation of the road is desirable. This is a particularly critical part of the route since clearly the Tamur River must be crossed, and in order to do this the road must wind down the steep, unstable valley sides. Hence, identifying stable areas of valley slopes at the river crossings is of major importance.

A typical generalised block diagram for one of the river valleys traversed by the road is shown in *Figure 7.17*. It shows a three-cycle landscape with fairly gentle upper valley sides, steep mid-valley slopes with scree slopes and degraded landslide scars, and finally an incised gorge section showing almost continuous instability. From this generalised picture of the geomorphology in *Figure 7.17* it is readily apparent that the most stable ground for route location is on the upper slopes. The gorge slopes are continuously unstable and too steep, and the mid-slopes are locally unstable and threatened by

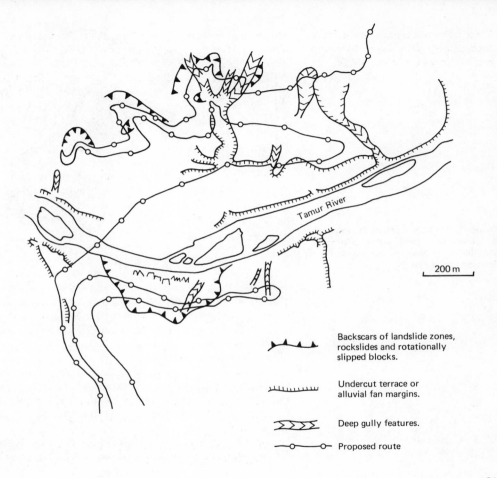

Tamur River

200 m

Backscars of landslide zones, rockslides and rotationally slipped blocks.

Undercut terrace or alluvial fan margins.

Deep gully features.

Proposed route

Figure 7.16 A simplified version of a geomorphic map prepared for the Dharan–Dhankuta road project (After Brunsden et al., 1975)

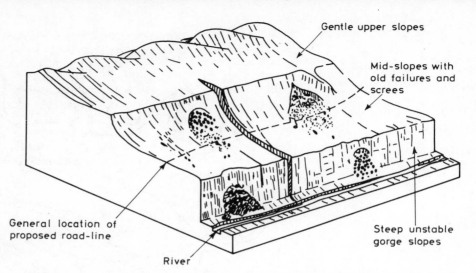

Gentle upper slopes

Mid-slopes with old failures and screes

Steep unstable gorge slopes

General location of proposed road-line

River

Figure 7.17 Block diagram to illustrate the general geomorphology over part of the Dharan–Dhankuta road project (After Brunsden et al., 1975)

instabilities occurring on the gorge slopes below. A general recommendation from the mapping could therefore be made, that a substantial part of the route should be relocated onto higher ground. Height should be made quickly in a series of hairpin bends over the stable parts of the valley sides, and above all, long crossings of the unstable midslopes should be avoided.

This project shows clearly that very severe difficulties can be encountered in civil engineering projects on natural slopes. At the same time, it demonstrates that a full appreciation of the form and processes of the natural slope allows problems to be minimised or avoided completely. In short, it is essential for engineering to be sympathetic to the natural landscape so that projects can harmonise as completely as possible.

7.3.3 Conclusion

Throughout this book we have taken the view that hillslopes are an extremely important part of the academic discipline of geomorphology, and arguably the most important since the whole of the landscape is composed of slopes subjected to processes of sediment and solute movement. We consider these processes to be of prime importance in determining slope morphology, and hence we have paid a lot of attention to the way they operate.

But also throughout this book, and especially in this chapter and Chapter 1, we have stressed that man's activities on hillslopes cannot be divorced from their natural geology and geomorphology. Consequently, the interactions between geomorphology and applied civil engineering are strong and two-way. On the one hand geomorphology aids successful use of hillslopes to man's advantage, while on the other hand civil and agricultural engineering have provided an enormous input to geomorphology which has enabled the discipline to progress very much faster than would have been possible otherwise.

Hopefully the few examples of agricultural and engineering problems described in this chapter have helped to highlight the importance of an understanding of hillslopes to man. The geomorphic complexities of the natural landscape mean that every project on slopes must be regarded as unique, and therefore deserving of special, detailed attention.

GLOSSARY

AGGRADATION	The raising of the level of a floodplain and river channel by progressive deposition of sediment (see *degradation*).
ANION	Negatively charged atom or molecule (see *ion*).
AQUIFER	Body of rock capable of holding and transmitting water.
ARTESIAN CONDITIONS	Rising of water in a well or spring by hydraulic pressure.
BASAL EROSION	Erosion concentrated at the foot of a slope.
BASE LEVEL	Limiting level of effective stream erosion (sea level or a locally-controlled level, such as a lake).
BEDDING PLANES	Structural weaknesses in a rock formed at time of deposition and separating individual beds.
BIOSPHERE	The total of all living organisms and their environment.
CAMBERING	Downslope collapse and subsidence of blocks of a hard rock unit due to deformation in an underlying weaker stratum.
CATION	Positively charged atom or molecule (see *ion*).
CATION EXCHANGE	Exchange of one cation for another upon a mineral surface.
CLINOMETER	Instrument for measuring slope angle.
COLLOIDS	Particles of extremely small size, capable of remaining indefinitely in suspension.
DEGRADATION	Lowering of floodplain and river channel by progressive erosion (see *aggradation*).
DENUDATION	Action of all processes to weather and transport rock materials.
DISCONTINUITY	Plane of structural weakness within a rock (see *bedding plane*; *joint*).
EDAPHIC	Relating to soil from point of view of plant growth.
ELUVIATION	Transfer of sediment vertically through soil profile in percolating water.

ENDOGENE	Relating to subsurface actions.
EPIGENE	Relating to surface actions.
EQUILIBRIUM FORM	Form developed under sustained action of a process, with steady and uniform surface lowering.
ERODIBILITY	Resistance of the soil to detachment and transport.
EROSIVITY	Measure of the power of rainfall to cause soil erosion.
ESCARPMENT	Landform due to erosion of a low to moderately dipping bed of resistant rock, forming an asymmetrical ridge with a steep (scarp) face and gentle (dip) slope.
EVAPORATION	Process in which water passes from liquid or solid state to gaseous state.
FAILURE	Point at which strength of material is exceeded by applied stress.
FORCE	Mass multiplied by acceleration.
GRADED STREAM	Stream whose long profile is adjusted so that average bed load transport is equal to input of bedload to channel.
GRADIENT	Tangent of slope angle.
HARD PAN	Accumulation of leached dissolved material, redeposited as cemented layer at depth in soil (see *leaching*).
HEAD DEPOSITS	General term describing material deposited by gravitational processes on slopes.
HYDROLOGICAL CYCLE	Movement, exchange and storage of earth's water in all states.
HYDROLOGY	Science of earth's water and its movements through the hydrological cycle.
INFILTRATION	Vertical movement of water into soil during and after rainfall.
ION	Electrically charged atom or molecule, due to addition or removal of electrons.
JOINT	Structural weakness in rock formed subsequent to deposition and involving little or no displacement.
LATERAL EROSION	Erosion due to lateral shift of a river at base of slope.
LEACHING	Removal of solutes from soil horizon by infiltrating rainwater.

LEVÉES	Low ridge of sediment alongside river channel formed by deposition during overbank flooding. Also applied to low banks pushed out on either side of debris flow to form trail.
LITHOSPHERE	General term applied to entire solid earth realm.
MORPHOLOGY	Shape of earth's surface.
PERMAFROST	Permanently frozen ground.
PODSOL	Soil type characterised by intense leaching and eluviation to produce iron- and clay-poor upper horizon and enriched lower horizon by redeposition.
POREWATER	Water contained in pore spaces of soil or rock.
RADIOCARBON DATING	Absolute dating technique based on rate of radioactive decay of ^{14}C within organic material and carbonates through time.
RECTILINEAR	Constant-angled or straight (slope).
REGOLITH	Layer of loose, uncemented sediment over bedrock derived from *in situ* weathering or transported from elsewhere.
RESURGENCE	Reappearance of stream from cave system (usually in limestone regions; see *sink*).
RILL	Channels eroded by flowing water on slopes during and immediately after rainfall. Usually seasonally destroyed.
RUNOFF	Flow of water from land surface in overland flow and streamflow.
SINK	Disappearance of stream into cave system (usually in limestone region; see *resurgence*).
SOIL HORIZON	Layer of soil within soil profile formed by vertical removal or enrichment by soil processes (see *soil profile*).
SOIL PROFILE	Vertical section of soil from surface through all horizons to parent material from which it was formed.
STRAIN	Deformation due to applied stress.
STRESS	Force per unit area.

SUBAERIAL	Existing or taking place on the land surface, exposed to the atmosphere.
TECTONIC ACTIVITY	Processes of folding and shearing of earth's crust.
TRANSPORT-LIMITED	Restricted or controlled by the rate of sediment or solute transport from a site.
WEATHERING-LIMITED	Restricted or controlled by the rate of rock weathering at a site.

REFERENCES

Allan, A.R. (1968). 'Coastal protection and stabilization at Herne Bay.' *Civil Engineering and Public Works Review* **63**, 860–861

Bayley, M.J. (1972). 'Cliff stability at Herne Bay.' *Civil Engineering and Public Works Review* **67**, 788–793

Bennett, H.H. (1939). *Soil Conservation*. New York, McGraw-Hill

Bishop, A.W. (1973). 'Stability of tips and spoil heaps.' *Quarterly Journal of Engineering Geology* **6,** 335–376

Brady, N.C. (1974). *The Nature and Properties of Soils*, 8th Edn. New York, Macmillan

Briggs, D. (1977). *Sediments*. London, Butterworths

Brunsden, D., Doornkamp, J.C., Fookes, P.G., Jones, D.K.C. and Kelly, J.M.H. (1975). 'Large scale geomorphological mapping and highway engineering design.' *Quarterly Journal of Engineering Geology* **8,** 227–254

Carson, M.A. and Kirkby, M.J. (1972). *Hillslope Form and Process*. Cambridge University Press

Carson, M.A. and Petley, D.J. (1970). 'The existence of threshold slopes in the denudation of the landscape.' *Transactions of the Institute of British Geographers* **49**, 71–95

Chandler, R.J. (1976). 'The history and stability of two Lias Clay slopes in the Upper Gwash Valley, Rutland.' *Philosophical Transactions of the Royal Society* **283A**, 463–497

Chandler, R.J. (1977). 'Back analysis techniques for slope stabilisation works: a case record.' *Géotechnique* **27**, 479–495

Finlayson, B.L. and Greenhalgh, M. (1969). 'A study of a mass movement in the Esk District.' *Capricornia* **5**, 62–66

Griggs, D. (1936). 'The factor of fatigue in rock exfoliation.' *Journal of Geology* **44**, 783–796

Hayes, W.A. (1977). 'Estimating water erosion in the field.' In *Soil Erosion: Prediction and Control*, National Conference on Soil Erosion, Purdue University, 1976. Arkeny, Iowa, Soil Conservation Society of America

Hutchinson, J.N. (1975). 'The response of London Clay cliffs to different rates of toe erosion.' *Building Research Establishment Current Paper* **CP 27/75**

Loughnan, F.C. (1969). *Chemical Weathering of the Silicate Minerals*. New York, Elsevier

Mason, B. and Berry, L.G. (1968). *Elements of Mineralogy*. San Francisco, Freeman

Mitchell, J.K. (1976). *Fundamentals of Soil Behaviour*. New York, Wiley

Mosley, M.P. (1973). 'Rainsplash and the convexity of badland divides.' *Zeitschrift für Geomorphologie, Supplementband* **18**, 10–25

Schumm, S.A. (1964). 'Seasonal variations of erosion rates and processes on hillslopes in Western Colorado.' *Zeitschrift für Geomorphologie, Supplementband* **5**, 215–237

Statham, I. (1973). 'Scree slope development under conditions of surface particle movement.' *Transactions of the Institute of British Geographers* **59**, 41–53

Statham, I. (1976). 'A scree slope rockfall model.' *Earth Surface Processes* **1**, 43–62

Statham, I. (1976). 'Debris flows on vegetated screes in the Black Mountain, Carmarthenshire.' *Earth Surface Processes* **1**, 173–180

Statham, I. (1977). *Earth Surface Sediment Transport*. Contemporary Problems in Geography Series, Oxford, Clarendon Press

Symons, I.F. and Booth, A.I. (1971). 'Investigations of the stability of earthwork construction on the original line of the Sevenoaks by-pass, Kent.' *Road Research Laboratory Report* **LR 393**

Whipkey, R.Z. (1965). 'Subsurface storm flow from forested catchments.' *Bulletin of the International Association for Scientific Hydrology* **10**, 74–85

Wischmeier, W.H. and Smith, D.D. (1965). *Predicting Rainfall Erosion from Cropland East of the Rocky Mountains*. US Department of Agriculture, Washington D.C., Agriculture Handbook No. 282

FURTHER READING

The following texts and articles include a number which are at an advanced level, only certain sections of which may be found useful.

GENERAL

Briggs, D. (1977). *Soils*. London, Butterworth

Carson, M.A. (1971). *The Mechanics of Erosion*. London, Pion Press

Carson, M.A. and Kirkby, M.J. (1972). *Hillslope Form and Process*. Cambridge University Press

Knapp, B.J. (1979). *Elements of Geographical Hydrology*. London, Allen and Unwin

Statham, I. (1977). *Earth Surface Sediment Transport*. Contemporary Problems in Geography Series, Oxford, Clarendon Press

Young, A. (1972). *Slopes*. Edinburgh, Oliver and Boyd

CHAPTER 1

Hails, J.R. (Editor). (1977). *Applied Geomorphology*. Amsterdam, Elsevier

Hudson, N. (1971). *Soil Conservation*. London, Batsford

Schumm, S.A. (1966). 'Development and evolution of hillslopes.' *Journal of Geological Education* **14(3)**, 98–104

Weyman, D. and Weyman, V. (1977). *Landscape Processes: An Introduction to Geomorphology*. London, Allen and Unwin

CHAPTER 2

Young, A. (1972). *Slopes*. Edinburgh, Oliver and Boyd

CHAPTER 3

Davidson, D.A. (1978). *Science for Physical Geographers*. London, Edward Arnold

Paton, T.R. (1978). *The Formation of Soil Material*. London, Allen and Unwin

Statham, I. (1977). *Earth Surface Sediment Transport*. Oxford, Clarendon Press

Whalley, W.B. (1976). *Properties of Materials and Geomorphological Explanation*. Oxford University Press

CHAPTER 4

Brunsden, D. (1974). 'The degradation of a coastal slope, Dorset, England.' *Institute of British Geographers Special Publication No. 7*

Burkalow, A. van (1945). 'Angle of repose and sliding friction — an experimental study.' *Bulletin of the Geological Society of America* **56**, 669–708

Chandler, R.J. (1977). 'The application of soil mechanics methods to the study of slopes.' In *Applied Geomorphology* (Ed. by J.R. Hails), pp. 157–182. Amsterdam, Elsevier

Prior, D.B. and Stephens, N. (1972). 'Some movement patterns of temperate mudflows: examples from north eastern Ireland.' *Bulletin of the Geological Society of America* **83**, 2533–2544

Terzaghi, K. (1962). 'The stability of steep slopes on hard, unweathered rocks.' *Géotechnique* **12**, 251–270

Young, A. (1960). 'Soil movement by denudational processes on slopes.' *Nature* **188(4745)**, 120–122

CHAPTER 5

Ellison, W.D. (1948). 'Erosion by raindrop.' *Scientific American*, Offprint 817

Finlayson, B.L. (1979). 'Electrical conductivity: a useful technique in teaching geomorphology.' *Journal of Geography in Higher Education* **3(2)**

Haigh, M.J. (1977). 'The use of erosion pins in the study of slope evolution.' *British Geomorphological Research Group, Technical Bulletin No. 18*

Hewlett, J.D. and Nutter, W.L. (1969). *An Outline of Forest Hydrology*. Athens, University of Georgia Press

Kirkby, A. and Kirkby, M.J. (1974). 'Surface wash at the semi-arid break in slope.' *Zeitschrift für Geomorphologie, Supplementband* **21**, 151–176

Ruxton, B.P. (1967). 'Slopewash under a mature primary rainforest in N. Papua.' In *Landform Examples in Australia and New Guinea* (Ed. by J.N. Jennings and J.A. Mabbutt), Ch. 5. Canberra, ANU Press

Schumm, S.A. (1956). 'Evolution of drainage systems and slopes in badlands at Perth Amboy, New Jersey.' *Bulletin of the Geological Society of America* **67**, 597–646

CHAPTER 6

Carson, M.A. and Petley, D.J. (1970). 'The existence of threshold slopes in the denudation of the landscape.' *Transactions of the Institute of British Geographers* **49**, 71–95

Chandler, R.J. (1976). 'The history and stability of two Lias Clay slopes in the Upper Gwash Valley, Rutland.' *Philosophical Transactions of the Royal Society* **283A**, 463–497

Gilbert, G.K. (1909). 'The convexity of hillslopes.' *Journal of Geology* **17**, 344–351

Hutchinson, J.N. (1975). 'The response of London Clay cliffs to different rates of toe erosion.' *Building Research Establishment Current Paper* **CP 27/75**

Mosley, M.P. (1973). 'Rainsplash and the convexity of badland devides.' *Zeitschrift für Geomorphologie, Supplementband* **18**, 10–25

Statham, I. (1973). 'Scree slope development under conditions of surface particle movement.' *Transactions of the Institute of British Geographers* **59**, 41–53

Wentworth, C.K. (1943). 'Soil avalanches on Oahu-Hawaii.' *Bulletin of the Geological Society of America* **54**, 53–64

Young, A. (1972). *Slopes*. Edinburgh, Oliver and Boyd

CHAPTER 7

Bennett, H.H. (1939). *Soil Conservation*. New York, McGraw-Hill

Hails, J.R. (Editor) (1977). *Applied Geomorphology*. Amsterdam, Elsevier

Morgan, R.P.C. (1979). *Soil Erosion*. London, Longmans

INDEX